TECHNOLOGIES OF POWER IN THE VICTORIAN PERIOD

TECHNOLOGIES OF POWER IN THE VICTORIAN PERIOD

Print Culture, Human Labor, and New Modes of Critique in Charles Dickens's *Hard Times*, Charlotte Brontë's *Shirley*, and George Eliot's *Felix Holt*

JOHN C. MURRAY

CAMBRIA PRESS

Amherst, New York

Requests for permission should be directed to:
permissions@cambriapress.com, or mailed to:
Cambria Press
20 Northpointe Parkway, Suite 188
Amherst, NY 14228

Library of Congress Cataloging-in-Publication Data

Murray, John Condon.
Technologies of power in the Victorian period : print culture, human labor, and new modes of critique in Charles Dickens's Hard times, Charlotte Brontë's Shirley, and George Eliot's Felix Holt / John Condon Murray.
p. cm.
Includes bibliographical references and index.
ISBN 978-1-60497-668-7 (alk. paper)
1. English fiction—19th century—History and criticism. 2. Speech in literature. 3. Dickens, Charles, 1812–1870. Hard times. 4. Brontë, Charlotte, 1816–1855. Shirley. 5. Eliot, George, 1819–1880. Felix Holt, the radical. 6. Communication and technology—Great Britain—History—19th century. 7. Literature and technology—Great Britain—History—19th century. 8. English literature—Publishing—Great Britain—History—19th century. 9. Books and reading—Great Britain—History—19th century. I. Title.

PR878.S64M87 2010
823'.809355—dc22

2009048868

To Allegra

Table of Contents

Acknowledgments

I would like to take this opportunity to extend my sincere thanks and heartfelt appreciation for the incisive and creative suggestions offered by my former advisor, Josie Campbell, since this book was developed from research that was based on my dissertation. I would also like to thank John Leo, Stephen Barber, Susan de los Heros, and Mario Trubiano for their contributions to my project and for serving as critical readers by making me examine and defend my assumptions.

Seldom do I have the opportunity to thank my students and colleagues who have offered help by challenging me or simply by taking the time to listen to me work with and through an idea. For all of their help, I extend my gratitude.

The staff at Cambria Press (especially Paul Richardson, Toni Tan, and Nicole Ford) was encouraging, helpful, and constructive in moving the book toward publication.

My sister, Colleen, was often considerate in keeping track of this diligent scholar. My parents, Martha and Jack, have been caring, understanding, and patient with the many leaps I have taken in my research and my life goals. Thankfully, and with much luck, I keep landing on my feet.

My fiancée, Allegra, has supported me loyally throughout the past few years on this project, as she has on everything I have ever undertaken. I cannot thank her enough for her love, patience, and for all of those small, tender moments that matter so very much. It is to her that I dedicate this book.

TECHNOLOGIES OF POWER IN THE VICTORIAN PERIOD

Chapter 1

Introduction

> Spoken language is not so much a thing that a person uses as it is a representation of the way a person is. Speech is to the mind as skin is to the body. It is the way a person comes to be a definite and expressive creature. Speech is as inseparable (even though distinguishable) from a person's thoughts and feelings as skin is from bones and muscles. And just as bodily movements are fluid, passing, and largely instinctive, so is spoken language.
>
> —Albert Borgmann

The purpose of my analysis is to employ Charlotte Brontë's *Shirley* (1849), Charles Dickens's *Hard Times* (1854), and George Eliot's *Felix Holt* (1866) to evidence how the growth of capitalist production and the development of new technologies of industry within the early to mid-Victorian periods inspired the prioritization of the printed word over oratory and speech as a means for fulfilling the linguistic power exchanges found commonly in spoken discourse.[1] Inventions such as Friedrich Gottlob Koenig and Andreas Friedrich Bauer's high-speed printing press enabled mass production and low-cost readership among the working class, who experienced literacy on multiple levels: to educate themselves, to experience leisure and diversion, to confirm their religious beliefs, and to improve their labor skills. Much in the same ways that speech had been used to affirm intersubjectivity, print culture conditioned readers to accept the unidirectional exchange of values and interests that would create a community of readers who would be responsive to the expansion of a new technical society and would eventually perform the routines of mechanized labor. Rather than merely romanticizing pretechnological cultures, I wish to suggest that the emergence of technologies of production and print culture within the early to mid-Victorian periods precipitated the diminution of linguistic exchanges as techné or modes of revealing and critiquing transferences of power and also for rivaling print culture's representational claims of how linguistic exchanges had been conceptualized and experienced.

The diffusion of print culture emphasized the "distinct advantage of writing" for mass production and dissemination. The implementation of new technologies made literacy an essential component of economic growth and was instrumental "in the making [of] a sophisticated work force that could understand and manage the new machinery and its operatives" ("Idea of Literacy"). During the early to mid-Victorian periods, industrial novels assumed the position of superiority in unacknowledged and unequal power exchanges with the reader as its "non-contradictory subject" (Belsey 60).[2] Novelists legitimized and authenticated narrative voice as a strategy for managing nineteenth-century dissatisfaction with fictional representations of linguistic power exchanges.

Early to mid-nineteenth-century Britain did not develop a public educational system to provide solutions for educating the working class. Cheap literature of the period served as a means for the working classes to locate themselves within society. However, the literature of the working class was less a reflection of society and more an instrument for regimenting the behavior of labor and for filtering out self-consciousness from the means and consequences of production. Industrial novels became textual supplements of reality that did not adequately provide substitutes for nonverbal cues used for interpreting figurative meanings, which became vague and susceptible to literal and factual representation.

In the preindustrial age, reading was a mode of social discourse. Through storytelling, memories existed through speech and were represented through the performative act of narration. Through the recitation of stories, individuals acquired a sense of communality, which was itself limited by the choices of "cultural material" and the "number of narratives" available for circulation (Fluck 24). Imagination was unfettered by individual desire and personal interest; rather, speech was used to affirm the traditions of oral culture. Improvements in print technologies enabled mass production in the early eighteenth century, but it was the emergence of the novel as a mode of literary, historical, social, and cultural expression that facilitated mass literacy during the early to mid-nineteenth century and inspired movement away from oratory and toward the private act of silent reading. The availability of literature, once the possession of the privileged few, for middle- and lower-class readers inspired a new literate culture predicated on silent reading, but even novelists who depended so readily on the written word could not avoid writing voice into their works or even engaging in direct dialogue with their readers through the use of narrative voice. It is through the performative act of speech that individuals contribute a verbal strand to the cultural narrative and bridge the gap between personal memory and cultural history.

Social theorists such as Winifreid Fluck suggest the "instrumentalization of print culture resulted in a remarkable gain in individual control," allowing the individual to "select material that comes more closely to [his or] her own imaginary longings and emotional needs" (24). Although Fluck is correct in suggesting that novels gave readers new modes of self-expression, she fails to acknowledge the ways in which readers may become susceptible to the limitations of mass literacy, which reduces if not homogenizes narratives to suit the demands of popular culture. The Victorian reader, removed from the public sphere, became responsive to fictional renderings by challenging their representation of the world in actuality. In moments of contemplation, Victorian readers considered the fictional realities of industrial novels that reflected contemporary social conditions, but by indulging in silent reading, Victorian readers secured privacy, insulated themselves from accountability, and fragmented cultural memory into disparate, competing narratives. Conceptions of Victorian readership emphasized the ways in which serial publication forced novelists to anticipate and to accommodate the demands of a responsive audience. Rather than continuously revising plots and characters to satisfy their readership, novelists engendered radical characters as a means for confronting social issues and for questioning their readers' commitment to reform. Radical characters such as Shirley Keeldar and Felix Holt enable the novelists to appeal to their audience while also stimulating critique and advocating a transformation from a homogeneous readership to a heterogeneous readership, a readership better suited for questioning the effects of industrialization.

Industrial novels are not, of course, pure representations of Victorian reality. However, many working-class Victorians engaged texts as authentic representations of society. How working-class readers then reconstructed their personal narratives in actuality suggests the effects of social assimilation upon subjective identity and advances the claim that industrial novels did not provide solutions to the social and economic maladies they reported. Rather, they contextualized social and cultural problems without recognizing the dangers of how the decontextualized imagination of the reader locates placement within the same ontological and epistemological assumptions.

As print culture asserted its representational claims, people who were being disenfranchised and shut out of the dialogue (i.e., the working class) developed new modes of speech through participation within crowd formations. Crowds rivaled the novels' representational claims by demonstrating their potential for envisioning new and distinct social orders yet to be realized (Plotz 2). Were it not for the aggregation of individual speech, crowds would be nothing more than a sign of themselves, a symbolic representation of speechless masses that inhabit the periphery of industrial novels. The ways in which speech is rendered or withheld within the crowd demonstrate its contradictions as well as its capacity for yielding linguistic exchanges. However, unlike the novel, the crowd's contradictions assume a physical embodiment that makes linguistic power exchanges possible.

Critics such as Nancy Armstrong and Garrett Stewart have considered the ways in which Victorian novelists such as Charlotte Brontë, Charles Dickens, and George Eliot addressed representations of speech in fictional discourse. However, these critics have not thoroughly evidenced "the simple fact that voice persists in the discourse of print culture, where it remains as trace and residue capable of giving rise to inchoate new forms" of speech, forms which might reengage critique of the human condition (Kreilkamp 2). In each of the selected industrial novels, the printed word does "turn on itself and invest extraordinary value in an idealized version of the speech community it had relegated to the past, a community for which the novel offered itself as both substitute and cultural memory" (Kreilkamp 3). It is this same community that has been either displaced by technical progress or relocated to the periphery of Victorian society as an anomaly, such as Sleary's circus in *Hard Times*. At the circus, the performers' manipulation of speech through creative uses of gesture, intonation, and other visual and auditory cues demonstrates how language games may be used as techniques of resistance to cultural narratives. It is through language games that Sleary's performers revalue how power is made equitable through linguistic exchanges. By allocating social spaces within society for anomalies like Sleary's circus, society becomes aware of its own structural deficiencies and falsely represents those deficiencies, a process that begins through the recognition of anomalous cultural formations that cannot ever be fully purged from within dominant culture and ends with the creation of fictional representations that prevent full ideological disclosure.

Each of the industrial novelists employs speech in different ways. Brontë uses speech as a signifier of individuality and collectivity within distinct types of crowd formations (i.e., passive or aggressive). Dickens stresses speech's potential for reconstituting ethical language and moral consciousness, for expressing alteration and contextuality, for exercising free will, and for making choices of accountability. And finally, Eliot offers speech as a reminder of preindustrial society, in which customs, traditions, and rituals were orally performed as a means for reaffirming one's sense of belonging. In each case, there is a common theme of how linguistic power exchanges had been either altered or inhibited by technical progress and the diffusion of print culture. This alteration is further demonstrated by the ways in which characters occupy conversational spaces to experience power exchanges from within not only their fictional realities but also the literature they choose to read or are instructed to read.

This study will examine and evaluate the ways in which the selected novelists offer readers middle-class, radical, social antagonists as metaphors of critical speech. These characters challenge both representations of linguistic power exchanges in early to mid-Victorian society and the reliance "on a form of power that works through language—and particularly the printed word—to constitute subjectivity" and "[to] perpetuate hegemony" (Armstrong 25). The epistemological paradoxes between speech and the printed word are emphasized by the ways in which Victorian culture valued oratory and speech as a strategy for advancing a social critique of the human condition. Through public speaking, listeners benefited from the information and emotion that was embodied by the orator while also recognizing the immediate value of competition, contradiction, and contradistinction that speech allows, specifically, the opportunity to stimulate immediate critical conversation. Each of the novelists addresses, either directly or indirectly, the failure of print culture to satisfy the needs of linguistic power exchanges, and thus, each novel devises new modes of speech that might reconfigure how such exchanges may be experienced anew.

The selected novels need to be situated within their historical context, from the Luddite riots of 1811 to the First Reform Bill in 1832. To this end, I employ historical methods laid out in Matthew Arnold's *Culture and Anarchy*, Pierre Bourdieu's "On Symbolic Power," and Mikhail Bakhtin's *Rabelais and His World* and Ernst Bloch's concept of simultaneity as points of critical intervention into the novels. While the effects of technologies of production on early to mid-Victorian society have been analyzed and evaluated by historians, economists, sociologists, and literary theorists, the ways in which these technologies have been employed as a means for valuing culture and critique provide further points of intervention.

Technology is a term derived from the Greek word *technologia.* Its etymology can be traced to Aristotle and its meaning revealed by the concatenation of *techné* ("craft" or "art") and *logia* ("saying") as the systematic treatment of art. Aristotle's definition may be applied to the ways in which technology has affected the human capacity for performing critical speech. Speech, suggests Aristotle, provides proof of character (Aristotle, *Rhetoric* 1.2.1356a–1358b). It is through speech that one engenders character and, thus, fulfills one's commitment to the production of human labor, which involves not only the production of goods and the fulfillment of services but also the creation and perpetuation of social relations. Ancient Greek civilization prioritized oratory as a means for engaging social criticism and inspiring individual accountability within public forums. Through public debate, listeners benefited from the information conveyed by the orator while also recognizing the value of argumentation that speech allows, specifically, the opportunity to stimulate critical conversation through contradiction and contradistinction.

Ancient Greek civilization was an oral culture, and writing assumed a tertiary status, unlike postmodern society, which emphasizes technologies of writing. Ancient Greeks used rhetoric as a measure of ethos, or character. Aristotle emphasizes how orators were able to demonstrate their linguistic competencies and individual character through traits of knowledge (*phronesis*), virtue (*arete*), and responsibility to one's audience (*eunoia*) (*Rhetoric* 2.1.137a). Public deliberation also allowed orators the opportunity to affirm and to represent not only their individual identity and agency but also the collective identity and historical agency of their community. Each rhetorical situation provided listeners with opportunities to reevaluate orators' virtues through their proximity to truth claims. Greek orators revealed special knowledge of issues affecting the community while reminding their audience of their own obligation to become responsive to the issues at hand. When speech was rendered for public consumption, Greek orators were scrutinized in terms of whether normative values and expectations were being satisfied or left unfulfilled by their deliberations.

The communicative symmetry of the Greek oratory model provides a working example of the ideal speech situation for rationalizing consensus and is a point of intervention into critical and historical debates of how radical antagonists found in industrial novels indirectly shaped public opinion and expressed cultural and social interests within the bourgeois public sphere of the nineteenth century, as well as demonstrated its transition from oral culture to literate culture. In the modern and postmodern ages, the written word has been further prioritized over oral traditions, which have not been so easily replicated or mass-produced. Although textual reproductions convey properties of language such as grammar and syntax, they fail to adequately reproduce sentiments, gestures, and intonations that empower language and make speeches memorable and transcendent.

The growth of the Internet in the late twentieth century and the emergence of a new technological sphere have been limiting the potential for critical speech as an essential component of communicative rationality. Digital communication has yet to fully achieve its potential for invigorating social exchanges of critical thinking. Rather, the Internet has been inspiring passive reliance on technical processes that have altered the consumption and production of information once used to formulate critical responses within the public sphere.

Instrumentalized modes of information and communication have been modifying how knowledge is exchanged and valued. However, the presence of instrumentalized modes of discourse and their effects on language are not new concepts. Yet, they have been erroneously historicized as by-products of modernization rather than as perpetual effects of industrialization during the early and middle Victorian era. The encroachment of technology into the Victorian public sphere transformed how people communicated in the mid-nineteenth century, a transformation realized by the diminution of critical speech. The displacement of subjectively informed communicative actions within objectively framed, technically oriented systems of reasoning inculcated apathy by demonizing critical speech.

I do not wish to imply that technology has not advanced human interests, for it has done so without hesitation. Nor do I wish to argue that technology may not again be called in service of such interests. However, I do hope to explain how and why technical progress has been obfuscating social concerns in order to assure the direct dissemination and implementation of technical-value rational imperatives throughout society. I do not intend to argue that technological achievements and human history maintained parallel courses that became divergent, for the error of such a conclusion would not require further evidence. Rather, I wish to suggest that human agency has assumed a tertiary status to technical processes and human history has abridged its narrative to suit the demands of technical progress. Rather than diverging from human history, technology has been subsuming it, and this process of subsumption was initiated during the Industrial Revolution and has continued to the present day.

In postmodern society, technologies of sign systems have been affecting conceptions of the self through the regulation of discourses of production and consumption. Technologies of sign systems have been transforming the ways in which knowledge is communicated and valued. Innovations such as the printing press modified how the written word was to be authenticated and disseminated. Technologies of print production disrupted the unity of subject by disconnecting the body from the writing process, much in the same way the manual laborer has been disconnected from capitalist modes of production.

Despite their varying availability to classes and genders, technologies of production became the constant reality of meaning against differentiated social subsystems and became ingrained in the Victorian cultural subconscious through the spread of the printed word. Technologies of production began to determine progress at their own rate and have since been assuring the fulfillment of internal functions. Technical progress has been measured as "the substitution of machinery, not merely for the skill of human hand, but for the relief of human intellect"[3] as well. For this reason, those who have employed normative and purposive categories for considering technologies of production and those who have argued that technologies of sign systems are a privileged medium that both prioritizes and reflects human interests have been failing to realize how technologies have rendered all of society accountable to their self-sustaining purposes.

During the early nineteenth century, industrialization facilitated new technologies of production that altered labor and communication by reconstituting sign systems to accommodate the demands of new technical processes. Sign systems inform concepts of reason and meaning. How we learn to recognize signs determines our conceptions of reality and our relation to the present culture. This process of signification has been complicated by intertextuality as well as the relation between reader and text and the culture they both signify and place under erasure.

Victorians embraced a logic of differentiation. Social subsystems of law, politics, art, science, education, and religion became self-referential modes of discourse supported by canons of literature. To facilitate this process of differentiation, Victorian society generated binary oppositions of self and other to determine whether normative values and expectations were being satisfied, especially within novels. By employing binary schemes as a means for regulating normative values and expectations, Victorian society enabled the possibility of technological intervention within the bourgeois public sphere. The organization and regulation of labor in the Victorian age, for instance, fixated on binary comparisons of productivity and nonproductivity which routinized tasks and valued technical objectives.

The manual laborer, who once had been able to fulfill the task of spinning several spools of cotton into yarn within a given time, was now able to increase factory productiveness within half the required time with the help of inventions like Reverend Edmund Cartwright's power loom, later perfected by William Horrocks and made more efficient through the use of James Watts's steam engines. Although such innovations accelerated the growth of capital, they also diminished the value of skilled manual laborers, who became insubstantial and interchangeable apparatuses of industrial-technical processes. Expunged from the factory system, the nonproductive laborer experienced separation from technical processes as loss of socialization. But, the presence of belonging had been nothing more than a simulacrum, an illusion of socialization legitimated by epistemes and a priori conditions which abstracted and reduced identity within the technical sphere. Concepts of belonging have been evolving into embodiments of class, gender, and physiognomy. Analyzing and evaluating the evolution of these conceptions through the selected novels of Dickens, Brontë, and Eliot reveal how these embodiments have been appropriated for technologies of production. These embodiments require a human presence in order to be perceived and expressed. However, technical intervention into sensory modalities has altered the process of human perception and the language used for social commentary.

Texts are not pure representations of reality, but they are constituent parts from which multiple layers of empirical and fictional reality are continuously formed at a constant rate within the mind of the reader. Regardless of their psychological demands upon the reader, the antagonisms between empirical and fictional reality rely on linguistic and nonlinguistic representations for their reconstitution. Much in the same way as we come to know ourselves in relation to what we are not, binary oppositions rely upon linguistic representation to reaffirm, to announce, and to critique the self. The transition from oral culture to literate culture caused slippage between how linguistic and nonlinguistic representations were created, experienced, and used to inform the relation of self and other.[4] The cultural alteration also anticipated the incompatibility of new technologies of production and radical critique within Victorian society. As a consequence of the transition to nonverbal modes of discourse, technologies of production instrumentalized Victorian conceptions of the relation of the self and the other in specialized and functional terms that predicted behaviors and established new boundaries for signifying meaning and identity in nonlinguistic forms.

When readers open a book, textual meanings undergo a perpetual process of revision and reconstitution. Through novels, for example, readers are able to trace the absence of reality from the fictional presence that informs it. Thus, texts are opened to the limitless possibilities of contextualization. Texts are not systems of enclosure limited to historical or literary periodization, even though they make possible that function. Rather, texts interact with the world, and the world responds in kind.

Victorian readers' constant movements of signification between exteriority and interiority framed the relationship between fictional and empirical reality and made true meaning reducible to the impermanence of signs always placed under erasure. Victorian readers were left to supplement texts and the world around them with their own productions of meaning, which also must be supplemented by an emergent technical language. However, "the supplement," Jacques Derrida explains, "occupies the middle point between total absence and total presence" (*Of Grammatology* 202). The supplementation of linguistic meaning by nonlinguistic forms merely fills an absence devoid of meaning, but the supplement does not achieve full plenitude of meaning without additional signs and supplements. As Derrida explicates,

> On the one hand, [supplementarity] is that of which the possibility must be thought before man, and outside of him. Man allows himself to be announced to himself after the fact of supplementarity, which is thus not an attribute—accidental or essential—of man. For on the other hand, supplementarity, which is nothing, neither a presence nor an absence, is neither a substance nor an essence of man. It is precisely the "play" of presence and absence, the opening of this "play" that no metaphysical or ontological concept can comprehend. … Man calls himself man only by drawing limits excluding his other from the play of supplementarity. (243)

Readers participate in a process whereby signs and signifiers are always being recontextualized and reinscribed. Although texts do reach conclusions, those conclusions might be revisited and contested through further examinations. Therefore, texts do not end at the last word and engage other texts, other readers, and other realities. These connections evidence how nothing fully exists either outside of or inside of the text without consideration of difference. Texts function in the absence of reality as "the immediate presence of content" (243), presence that requires further signs and supplementation which point to additional chains of signs and supplementation that defer, rather than assure, the fulfillment of meaning.

Texts are imbued with potential meanings of difference, but also are imbued with actions and consequences rooted in the empirical world. Texts are not merely typewritten records of language or supplements of oral discourse, but are narratives that enable reconstructions of meaning and identity. The black squares of print are not the boundaries of conceptualization, and their white margins are not the only spaces difference might occupy. The world and the reader as subject are also texts, and their presences exist both within and beyond their structural fixity. Through deferrals of meaning, the supplement signifies an "in-betweenness" of meaning and nonmeaning, of identity and nonidentity, of existence and nonexistence that expresses "inherent lack" of reified center (Derrida, *Of Grammatology* 88–89).

Decentering texts and destabilizing signs allow us to trace meanings of difference outside of textual and theoretical margins and enable us to partake in the unrestricted semantic play of limitless possibilities of interpretation. Derrida offers "two interpretations of interpretation, of structure, of sign, [and] of play." The first is the "lost or impossible presence of the absent origin" (*Writing and Difference* 292). The second is "the joyous affirmation of the play of the world and of the innocence of becoming, the affirmation of a world of signs without fault, without truth, and without origin" (292). Recognizing the possibilities of difference prevents the totalization of opinions by reengaging conceptions of contestation and contradiction. Theoretically, radical critique might be traced through the multiplication of supplements that seek deferral of meaning and might, therefore, be reconstituted through the process of differentiation.

However, technologies of production require consensus of principles of labor and capital. Human socialization within the public sphere has been contingent on obedience to and performance of technical principles that disable radical critique of labor and capital by interrupting the process of differentiation. Technology, therefore, has been becoming a total power that humans might neither eschew nor escape. Technologies of production, therefore, have also exerted control over the sign systems used to reveal meaning and reality. Technology, once a means for delineating class distinctions, has been becoming value-neutral because it has placed under erasure the ways in which differences have been privileged and signified. Technical rationality, therefore, has exerted control over values used to signify meaning and identity. If technology has achieved value bias, it is only by way of self-preservation and the perpetual signification of technical principles that are themselves situated within other forms of knowledge. To consider technology as nothing more than a different social subsystem would diminish how technologies of production have been successfully interwoven into other systems of knowledge, so much so that technology cannot be differentiated from them.

The regularity of technical processes has become so commonplace that the logic of differentiation has been eschewed in favor of automated labor that subsumes individual ambition and character within units of homogeneous production. Technical processes foster a sense of purposiveness, a sense of belonging, which obscures how technical rationality severs the links between mind and body, between thought and action, and between realization and resistance. The individual who confuses imaginary fears of ostracization with an actual lack of belonging is not able to symbolically represent this lack of belonging and is made powerless by its seeming absence. Without differentiation, without opposition, social criticism lacks the grounds for arguing for improvement. What has been lost to the measure of technical progress is the critical agent of human history: the social antagonist who mends the fissure between body and mind and speaks in the forgotten tongues of revolution and reformation.

Technologies have been employed by dominant social groupings to perpetuate class attitudes and habitus, classifications that restrict the potential for socioeconomic movement within society. The growth and expansion of technology into other disciplines necessitated new technical schemes of communication. Critical modes of interrogation and deliberation were replaced by concise, operational codes and patterns of behavior that expedited economic production. These codes and patterns were replicated through habit and evolved into technical principles of rationality internalized as attitudes and values that inform human interaction. By reinterpreting social legitimations, technical rationality purposefully designates new boundaries of normative order, boundaries which merely begin to be delineated within the Victorian age and are textually reproduced within the selected novels. Chapter 2 introduces readers to Dickens's *Hard Times*, in which the circus is represented as an organic counterculture that evolves in response to the values and traditions of social and industrial reengineering of the eighteenth and nineteenth centuries and patterns of labor and social behavior for more than one hundred years.

Sleary's circus, as counterculture, makes critique available for intervention into systems of technical rationality, systems that promote factual reverence and conceptual indifference.

This chapter depicts the circus as a heterogeneous group with a shared language, senses of purpose, and community. Their commonality parallels the ways in which nonproductive laborers experience separation from technical processes as a loss of socialization, a loss of language, and a loss of belonging within the factory system. Yet, the presence of belonging had been nothing more than a simulacrum, an illusion of socialization legitimated by systems of knowledge which have been abstracting and reducing identity within the technical sphere.

Chapter 3 considers the depiction of working-class crowds as both physical entities and theoretical abstractions of discourse in Brontë's *Shirley*. Analyses and evaluations of crowds in *Shirley* suggest how the objectification of crowds was employed by Victorian novelists to normalize patterns of human interaction. It also demonstrates how conditions of dynamism might be used to achieve movement between political radicalism and ideological conformity or between social aggression and civil passivity.

Since the early nineteenth century, concepts of belonging have been evolving into embodiments of class and gender. Chapter 4 analyzes and evaluates the evolution of these conceptions in Eliot's *Felix Holt* and considers how these embodiments have been appropriated for technologies of production. Although the embodiments require a human presence in order to be realized and performed, technical intervention into sensory modalities has altered the processes of human perception and expression and therefore reconfigured the ways in which critical language might be deployed.

The struggle to assert one's autonomy within a pluralistic society was one of the most prominent themes of early nineteenth-century literature. It not only imprinted itself on the literature of the period, it also reconstituted how conceptions of differing systems of knowledge must be reconciled in order to authenticate place-identity, which inscribes the relation of self to other by assuring status and location within society. Chapter 4 examines Eliot's novel *Felix Holt*, which presents society as an organism acted upon by historical forces that shape its contours and excite its evolution. Eliot maps out a world in which seemingly unrelated choices have an interconnectedness that renders each member of society a contributor to macroscopic consequences. The character of Felix Holt echoes the disposition of society and embodies the social and moral concerns of his time while searching for his place-identity in terms of the natural, cultural, and psychological landscapes that inform it.

Although this study references texts that fall under the rubric of both "social problem novels"[5] (*Shirley* and *Felix Holt*) and "industrial novels" (*Hard Times*), it analyzes and evaluates the ways in which these novels reflect how critical language has been lost to the process of technical replication. I also suggest the ways in which these novels might anticipate the loss of language as a symptom of industrialization and trace the absence of the critical voice through the performative act of speech. Each of the selected novels offers readers radical characters whose thinking eschews technical rationality in favor of communicative rationality. This work examines and evaluates the ways in which social antagonists suffer a loss of critical language and inevitably succumb to the technical principles advocated by more conservative characters. Felix Holt and Shirley Keeldar, for example, both struggle to trace the absence of speech as a lost techné of communicative rationality, while more conservative characters like Josiah Bounderby and Robert Moore signify how technical rationality overcame communicative rationality by limiting the potential for critique within the emergent technical sphere.

I use Jürgen Habermas's *Theory of Communicative Action* to analyze and evaluate how technical rationality was adopted within the dominant culture as a strategy for causing the absence of critical language from Victorian social discourse to maintain the socioeconomic status quo. The transformation from communicative rationality to technical rationality necessitated the renegotiation of British conceptions of how one occupies public and private domains by representing women and working-class characters as the embodiment of the nameless and faceless masses that populate the landscapes of industrial novels. This study employs Habermas's theoretical language to accommodate Victorian concepts of class and gender determinism within technical conceptions of noncoercive consensus.

In the study, I also explore the dynamic tension, found in selected industrial novels, between the ideas of technology as an autonomous and self-referential instrument of nonsocial rationality and technology as a neutral component that is responsive to socially determined interests of the Victoria era. The study locates historical attitudes and sentiments within the context of industrial novels to suggest the effects of technology on individualism and provincialism within the Victorian age as well as its transition from oral culture to literate culture. I consider how industrial novels anticipate the incompatibilities of technology and critique in the postmodern age by offering the social antagonist as a metaphor for retrieving the lost techné of public speech from the displaced knowledge of preindustrial culture.

Borrowing from Max Weber, Habermas's theory of rationality suggests the organizing feature of society is to shape knowledge into templates of rational apprehensions that become facts: "the world of systems … colonizing the world of life" (*Communicative Action* 81). However, Habermas's comments allude to the possibility of a suprasphere of technical rationality that subordinates all prior forms of knowledge and expression to the procedures of the technological sphere. While preindustrial culture anticipated the differentiation of multiple spheres of discourse, Victorian industrial society homogenized these spheres within an all-encompassing technological milieu.

Rethinking terms such as *efficiency*, *mechanization*, and *domination* might enable greater insight into the evolving relationship between "critical-oppositional intellectuals" or social antagonists (e.g., Felix Holt) who foster communicative rationality within the public sphere, and the technology that redefines how one may access and implement strategies of public discourse (Kellner). Finally, this study concludes by suggesting whether or not technology has become a supradisciplinary sphere of substantive rationality that enables human progress or whether or not it has appropriated formal rationality as a device for lessening the potential for social critique. By delineating and occupying technologically enclosed spaces for limited verbal communications,[6] humans are divested of substantive rationality and of an appreciation for procedural consensus. If two people agree to disagree, they are adhering to Habermas's prescription for noncoercive consensus. But, what have they have actually achieved? Consensus has been continuously employed as a sociopolitical strategy for achieving harmony and unity, but it has also been used to spread apathy and dissolution.

Consensus building has been used by dominant social groupings as a metaphor for maintaining socioeconomic status quos by limiting the potential for critical debate within the public sphere. However, not to pursue consensus neither necessitates nor promotes disharmony, disunity, or proactivity. Within the technological sphere, consensus building does not take place because man is no longer the subject of communicative interaction. Rather, man operates within the framework of technical principles and, thus, becomes the object of purposive rational action. In the public sphere, language has always been recognized as a man-made creation, continuously readapted to promote social exchanges. However, in the technological sphere, critical language collapses under its own theoretical weight and is left devoid of conceptual mass. Critique is not able to transcend the immanent plane of technological knowledge that comprises the technical sphere.

However, speech becomes the performative function of radicalism, a techné for reactivating the limitless possibilities of difference, for challenging the status quo by invigorating critique and releasing us from the limits of technical principles. A central pathology of industrialized societies has been the willingness to colonize the lifeworld by reconstructing and reinscribing it within interlocking systems of rationality, or, as Habermas terms it, "self-regulated systems of purposive-rational action and adaptive behavior."[7] Systems of rationality are employed by industrial societies to achieve dominion over natural resources that are used in service of social interests. The same systems of rationality are mobilized in support of individual or group interests and redirected against populations to regulate action and behavior by administering the production and distribution of technical and communicative resources.

Technology has been transforming societies, and some of the most exploitative, dehumanizing, and oppressive uses of new technologies have also been the most productive. During the Victorian era, British mercantilism employed technologies of industrialization to expand systems of trade by delving into international economic markets in order to promote agendas of imperialism and colonialism. The growth of mercantilism, in early nineteenth-century Britain, along with the emergence of the new industrial and technological sphere it expanded, limited the potential for engaging speech as the performative act of communicative rationality within Victorian life. Technology of power was advanced during this era. Inventions like the steam turbine altered the process by which power was generated and enabled the automation of factory labor, with dire consequences of unemployment and disenfranchisement for the assimilated and instrumentalized working class.

The integration of technology and manual labor during the Victorian age satisfied the demands of a primitive but growing economy that no longer required sufficient numbers of highly skilled workers to operate automated machines. As Victorian social theorist John Burnett suggests in his seminal work, *The Annals of Labour*, prior to the Industrial Revolution, workers were classified as either skilled craftsmen, "who had learnt a specialized craft or 'mystery' by apprenticeship," or unskilled laborers, "who had only muscular strength to sell" (56). However, the automation of production and the development of the "factory system"

> vastly complicated this traditional distinction by virtually creating a new category of manual worker—the machine-minder, or factory operative who, though intimately familiar with the particular, sub-divided operation of his machine, was able to exercise only limited judgement or discretion over its performance. For him—or for her, since a high proportion of such workers, especially in the textile factories, were female—a short period of training rather than a long apprenticeship was generally all that was necessary. (57)

Automated labor multiplied productivity by diminishing the time and resources needed to cultivate skilled craftsmen. Poorly educated workers were taught to adapt themselves to sequencing tasks that were sufficiently mimicked through repetition of technical processes. Carrying out technical processes required less individual thought and responsiveness and instead required the substitution of technology and capital for manual labor and management of production.

Minimally skilled laborers gradually and effectively replaced artisans and craftsmen, whose disenfranchisement supported the illusion of disembodied labor, of factories functioning independently of workers' individual attributes and characteristics. In effect, automation of labor should be considered a technological achievement with each instrumentalized human component made responsive to his or her minimal task but not made fully aware of his or her role in the larger project. The extraction of profit from automated labor enabled the accumulation of capital and the concentration of resources that hastened population swells within British industrial cities. What was made absent from this technical expansion was human labor which involved not only the production of goods and the fulfillment of services but also the creation and perpetuation of human relations, which were concealed and mediated by technical processes.

Within the factory environment, the desire to achieve maximum levels of efficiency and expediency displaced concerns of how automation might merge interpersonal communication and mass communication within a new and emergent technological sphere. New instrumentalized modes of information and communication, such as the printing press and electric telegraph, modified how human knowledge was to be transmitted or received, collected or distributed, altered or perpetuated, validated or denied. Although the possibility of communicating through electric currents was a commonly accepted means of communication in the Victorian era, its availability as a tool for mass consumption and dissemination has been considered highly questionable.

The presence of instrumentalized modes of discourse such as the electric telegraph encroached into the public sphere and led to the delinguistification of social interaction: that is, the displacement of subjectively informed communicative actions within objectively framed, technically oriented systems of reasoning that inculcated apathy by demonizing social critique. The written word was prioritized over oral traditions that were not able to be so easily replicated or mass produced. However, public speaking was regarded as not only an art form but also a performative function of membership within the public sphere. In light of Habermas's conceptions of technology as an "instrumental form of nonsocial rationality" ("Technology and Science" 105), it will be pertinent to tease out whether or not technologies of industrialization compromised and forever altered conceptions of the public sphere by limiting the potential for spoken discourse as a human techné,[8] a countermeasure to technical rationality.

Victorian culture emphasized the value of oratory and speech as an accepted technique for advancing the process of social critique. The diminution of social critique within the Victorian technological sphere caused the disruption of the repetition of rhetorical values of critical language which had been stabilized by the performative act of speech. On a deeper level, it also suggested how technical rationality has been used by dominant social groupings as a strategy for divesting spoken language of its voluminous possibilities for invigorating critical conversation.

What has been theorized is whether or not technology might be responsive to a new critical language. However, a more radical critique has not been proposed in order to delay the processes of technological domination, processes that might have been halted during the Victorian age but might now only be interrupted by postmodern, anticritical intellectual banality. It is, therefore, important to devise appropriate techniques for countering the effects of technological change, even if those techniques provide only respite. As Herbert Marcuse intimates,

> At the nodal points of the universe of public discourse, self-validating, analytical propositions appear which function like magic-ritual formulas. Hammered and re-hammered into the recipient's mind, they produce the effect of enclosing it within the circle of the conditions prescribed by the formula. (*One-Dimensional* 88)

Linguistic patterning common to speech may be the "formula" to which Marcuse is hinting.

There are three elements that are essential for critical conversation to be successful: semantic referents, speaker referents, and actual referents. Semantic referents offer meanings that are readily perceivable and commonly understood. The speaker's referent is, quite simply, "what is being said," and the actual referent is the context, or "what is actually meant" by the speaker. Although the speaker's utterances may differ somewhat from the listener's reception, the use of semantic referents allows for a more fluid interaction that assures the realization of actual meaning through the use of intersubjectively shared language.[9]

The symmetry between speech and speaker and sign and context demonstrates how embodied humans interact with their natural world. However, "this symmetry is thought to have split down the middle [and] the resulting division of mind and world seems to have given humans dominion over reality" (Borgmann 22). The division between mind and body, thought and expression, speech and writing experienced a transformation during the Victorian age as language was exposed to the demands of technology and yielded to technical discursive control.

Thought and inscription have been commonly linked as a singular, descriptive term. Thought-inscription reaffirms the value exchanges that designate subjectivity in relation to community. However, thought anticipates the differentiation between speech and inscription and initiates perpetual subvocalization of self-awareness. Thought, then, is subvocalized, anticipates speech, and is supplemented by the presence of inscription. What must be theorized is whether or not technology has been transforming the process of supplementation so much that speech and writing require new modes of completion that can no longer be achieved without technologies.

Technologies have not been merely supplementing writing and speech but have been altering and replacing them. What also must be considered is whether or not technologies might be responsive to a new critical language that might interrupt if not reverse the effects of technical supplementation. A more radical critique must be instigated in order to delay the processes of technological domination, processes that might have been halted during the Victorian age but might only be interrupted within the anticritical postmodern age. It is, therefore, important to devise appropriate linguistic techniques for countering the effects of technological change, even if those techniques provide only respite. Technologies function according to principles of efficiency and homogeneity; therefore, technologies have been becoming the "supplement of supplement" (Derrida, *Of Grammatology* 303) Technologies are Derrida's "originary supplement" (313), and technologies "[come] in the place of a lapse, a nonsignified or a nonrepresented, a nonpresence. There is no present before, it is not preceded by anything but itself, that is to say by another supplement" (303–304). The very act of supplementation leads to differentiation. The intervention of technologies within this process has been diminishing the potential for difference.

Within the industrial novel, the ways in which the separation between home and factory is made less distinguishable mirror how the separation of interior mind and exterior world is made less possible within a technological society. Technology bridges the chasm between conceptual and perceptual dimensions by creating a replicated world that does not distinguish between these realities. In light of such considerations, one must ask, "What, then, does verbal communication offer the characters of industrial novels that technology cannot replicate through the written word?" Also, one must question, "Does speech even exist once the words are uttered, heard, and forgotten?" And, if so, "Do texts disembody the performative act of speech, so much so that the reconstruction of words on a page does not allow for the smooth dissemination of intended meaning and the loss of intersubjectivity that enabled the navigation of the public sphere in preindustrial culture?"

What has been lost in this process of interpretation is the expressivity of self-awareness that links speakers (and readers as well) to the natural world rather than to supplemental reproductions that are textual replications of technical reality which eschew social and moral implications by prioritizing efficiency of information. Through repetition and habit, readers begin to weave man-made narratives into the tapestry of their surroundings. How readers then go about reconstructing their narratives in actuality suggests the effects of social assimilation upon subjective identity and advances the claim that industrial novels do not provide solutions to the social and economic maladies they report. Rather, they contextualize the problems without recognizing the dangers of how the decontextualized imagination of the reader locates placement within the same social network and epistemology of the novel.

The selected novelists embed functionalist and capitalist terms of reference within the context of their novels. The terms are used to precondition readers to operate within the technical framework of authors' observations—observations that authenticate technical rationality as an eventuality of human progress. However, from the Victorian age to the present day, history has not been measured by human activity but rather by instrumental progressions that lead to the abstraction of human attributes. For example, technologies of industry replicate human speech and model human capacity to process the content and context of knowledge and language. Verbal communications frequently employ linguistic devices or context cues such as pauses, intonations, facial expressions, and hand gestures to reveal intended meanings. Humor, sarcasm, figures of speech, and even silence are linguistic manifestations that coordinate points of reference for developing intersubjectivity. However, technology delinguistifies language by removing context cues from spoken discourse, which profoundly impacts the potential for occupying conversational spaces. What remains of spoken language is, to quote Martin Heidegger, nothing more than "the abstract form of writing that is transformed into the formulae of a logic," whose precision and clarity "ensures the rapid transformation of information" (*Überlieferte Sprache und Technische Sprache* 24).

Technologies have been both signifying and being signified by other things. For example, technologies of communication, like iPods and cellular phones, are signifiers of technological advancements that are themselves signified by the functions of the devices. Technology functions so normally and so habitually that it is quite difficult to regard it as merely a signifier of class status. When we now trace deferral or absence of meaning and reality, we do so through the presence of technology. As Derrida suggests,

> External/internal, image/reality, representation/presence, such is the old grid to which is given the task of outlining the domain of science. And of what science? Of a science that can no longer answer to the classical concept

> of episteme because the originality—an originality that it inaugurates—is that the opening of the "image" within it appears as the condition of "reality," a relationship that can no longer be thought with the simple difference and the uncompromising exteriority of "image" and "reality," of "outside" and "inside," of "appearance" and "essence." (*Of Grammatology* 33)

However, the absence of degrees of differentiation between technology and other systems of knowledge forces us to locate its presence both within its representation of those systems of knowledge and beyond them. Human language and history might eventually become, to borrow from Derrida, the representation of the originary unity and supplementation of technology. Technologies of production have been influencing the allocation and functions of status, procedural, and symbolic languages. Technology has become the sign (technology as representation) of what is being signified (technology undifferentiated from other systems of knowledge) by the signifier (technology as totalizing power).

Technologies of production diminish the possibilities of differentiation. The regularity of technical processes has become so commonplace that the logic of differentiation has been eschewed in favor of instrumentalized modes of labor which subsume individual ambition and character within units of homogeneous production. Technical processes foster a sense of purposiveness, a representation of belonging which obscures how technologies of production sever the links between mind and body, between thought and action, and between sign as the link between signifier and signified. Without differentiation, without alteration, social criticism lacks the grounds for arguing for improvement. Technical progress has eschewed the critical faculties of human language and history, which mend the fissure between body and mind, between thought and action, between sign, signifier, and signified, and reveals reality to us in the terms of alteration, differentiation, and opposition. Value positing enables the construction and perpetuation of human narratives of identity and history. The supplementation of moral and ethical values for technological imperatives abridges this process and brings human history to its conclusion.

Industrial novels evidence the incompatibilities of capitalism by suggesting the ways in which the absence of critical language causes the displacement of intersubjectivity by pluralizing what preindustrial culture regarded as a collective public sphere. Coordinated points of reference have been replaced by technical modes of intersubjectivity that have negated a priori and replaced communicative intersubjectivity which has been used for stabilizing links between distinct social, cultural, and economic locations in pretechnological societies. Capitalist modes of production permeated the Victorian age, creating a climate of contestation that limited the opportunity of consensus building by reducing and abstracting the working class and women.

The ways in which industrial novels compartmentalize members of disaffected groups within technical spheres ontologize degrees of ideological commonality and difference. Rather than pursuing resolutions to class and gender stratification, industrial novels perpetuate the status quo by allowing character-members of alternate spheres to interact through the exchange of technical principles and values that do not enable them to approximate their degrees of separation by distinguishing spheres of engagement through critique.

Habermas's concept of communicative rationality does not adequately explain why the Victorian technical sphere was not a place for rational-critical conversation, but rather a place where contestation and competition held sway over objective consensus building. Technological consciousness became the ideological nucleus of a Victorian order by eliminating the "distinction of distinction" between formal and substantive rationality ("Technology and Science" 113). Technologies of production no longer served as a means for advancing private interests, nor did they serve as devices activated in service of the public good. Rather, technologies of production assumed the character of domination and became the virtual embodiment of human consciousness. Spoken language, unlike its textual supplement, is not neutral and "[assumes] residence in a particular discourse, and alternative semantic systems or discourses are not just alternate views or versions of 'reality'" (Ermarth 3–4). The antagonistic relation between technical and communicative rationality may be redressed by reactivating critical thinking through the performative act of speech.

In *Nicomachean Ethics*, Aristotle writes,

> Every art [techné] is concerned with bringing something into being, and looks for technical and theoretical means of producing a thing which belongs to the category of possibility and the cause of which lies in the producer and not in what is produced. (See Book 6, *Nicomachean Ethics*, 88)

Speech, therefore, becomes the performative act of communication and the human embodiment of techné that has the potential for reactivating and reconstituting human critical concerns. For it is

> [t]he human body with all its heaviness and frailty [that] marks the origin of the coordinate space we inhabit. Just as in taking the measure of the universe this original point of our existence is unsurpassable, so in venturing beyond reality the standpoint of our body remains the inescapable pivot. (Borgmann 190)

Unlike technical thinking, speech as techné would be sublimated to serve the individual subject in ways that counter the effects of technical rationality.

One's inability to fully express and to rationalize the self limits the opportunity to critique or to adjust one's behavior in opposition to the possibility of coercion and domination of the self within the technological sphere. Thus, as Herbert Marcuse suggests, "Human behavior is outfitted with the rationality of the machine process, and this rationality has a definite social content" ("Social Implications" 144). The increase in adaptive behavior initiates, as Habermas suggests, "the dissolution of the sphere of linguistically mediated interaction by the structure of purposive-rational action" ("Technology and Science" 107). By imitating technical processes, however, human behavior is subordinated to the task at hand. Technical processes, therefore, regiment behavior of labor to filter out self-consciousness from the means of production, and what emerges from this cognitive cleansing is a techno-human apparatus devoid of critical thought.

Chapter 2

Exploring Carnivalesque in Dickens's *Hard Times*

> When they see to their horror how logic coils up at these boundaries and finally bites its own tail—suddenly the new fear of insight breaks through, tragic insight, which, merely to be endured, needs art as a protection and remedy.
>
> —Friedrich Nietzsche

Language, in all its many forms and functions, should enable the limitless possibilities of becoming. Through the alteration of language, cultures continuously reconstitute themselves by creating values and meaning that allow for deferral and revision. Yet, in the opening paragraph of *Hard Times*, Dickens's narrator announces the birth of a new language, a restrictive language that employs factuality as a strategy for regulating human behavior within Coketown by diminishing the possibility of differentiation. The precision of technical language limits the potential for expressing alteration and contextuality, for exercising free will, and for making ethical choices of accountability. Facts become the operative codes of a technical language bereft of moral consciousness and physical embodiment. It is through deferral and revision that technical language might be reduced, revalued, or rejected entirely. However, Dickens's novel provides contemporary readers a fictional representation of their own reality, as well as a reminder of the ways in which language might be deployed to simultaneously enable and inhibit communication and convey and withhold meaning.

Language has been a means for invigorating culture and meaning. Language is the signifier of culture, and it reveals cultural realities through human agency. However, technical language must not only produce itself but must also legitimate the ideologies of a technical utopia. Once legitimated, these ideologies become imbedded in the institutions of the society they seek to authenticate and assure cultural distantiation.

Culture is subordinated to the authority of these technical ideologies, and any attempts to resist them are merely signified as expressions of play or simulated experiences sanctioned by the same systems of knowledge they wish to usurp. Through alteration and differentiation, language has the potential for evolving and expanding the ways in which life experiences are communicated. Meanings may be engendered, revalued, and represented in terms which distinguish cultural milieus and redefine human identity. Without the possibility of differentiation, without the potential for alteration, language becomes static, meanings disconnect from their signifiers, and cultures become unsustainable. What is left in the absence of differentiation is the disembodiment of human agency within culture, the loss of interest in institutional reform, and individual disenfranchisement from modernity. Regarding this technical language as a movement toward progress and reform would be unjustifiable, especially since technologies offer false representations of human experience, representations that are continuously adjusted to diminish the power of human agency.

Within Dickens's novel, facts become the foundation of an emergent technical language, a language that has come to regulate not only the production of goods but also the potential for achieving differentiation. By stressing the primacy of factual language, Dickens prepares his readers for the imposition of technical principles upon all modes of human interaction. However, Dickens also was aware of the need for balance between dominant culture and counterculture, otherwise risking the violent reaction from within a society being overwhelmed by industrialization of labor and systemization of economic resources, something to which Dickens bore witness during the 1853 labor strikes and riots at Preston, a textile and manufacturing town in Lancashire.

Dickens provides the characters of Coketown with a mode of resistance to technologies of production advocated by Utilitarian characters such as Josiah Bounderby and Thomas Gradgrind, through education and literacy as ideological modes of liberation from one's human condition and Mr. Sleary's circus as a performative mode of resistance to social norms. At the circus, characters experience freedom from the constrictions of technical language. Their manipulation of speech through creative uses of gesture, intonation, and other visual and auditory cues demonstrates techniques of resistance to social narratives constituted through language games:

> The circus people's manner of speaking, with its private slang (outlandish to Gradgrind's ears), seems to emphasize the resistant material quality of language, rather than

> the communication of specific meanings. Sleary's heavy, bronchial speech does the same thing and reminds the reader incidentally of the "corpulent slow boy, with a wheezy manner of breathing" who mistakes Gradgrind's intentions in the classroom and offers free association instead of reasoning. (Dickens 1.2.8; Connor)

The imaginative language that facilitates the circus' presence, therefore, becomes the signifier of visible culture and of critical thought. It also demonstrates the ways in which language may be employed as a mode of manipulating and subverting meaning through constant deferral, a possibility that is negated within the systems of learning and labor in Coketown.

Subversion of the dominant culture is exemplified by the circus performers, whose transformations suggest the mutability of identity within counterculture. Kidderminster, for instance, is a caustic young man who performs the role of Cupid at Sleary's circus. In actuality, Kidderminster is boisterous, obnoxious, and temperamental, but in the middle of the circus ring, he is endearing and achieves spectators' approval. Through his imaginative performance, Kidderminster gains a measure of recognition, acceptance, and identity—even if these accomplishments are acquired through differentiation of deception, they evidence moments of liminality, when ethical choices might unravel the rigid social fabric of Coketown to instigate movement toward cultural heterogeneity.

To merely regard language as a system of grammatical rules and strategies for combining word choices devalues the complex nature of human interaction and the ways in which language is used to inform cultural and social identities. The contrasts between technologies of production and representations of culture are symbolized by Bounderby's factory and Sleary's circus, respectively, and suggest the ways in which language might be employed within different social spaces. The circus is more than spectacle. It becomes a venue for suspending social hierarchies and for disrupting the continuity of logic and factual reasoning. As Mikhail Bakhtin suggests, circus allows for the "temporary suspension of all hierarchic distinctions and barriers among men and of the prohibitions of usual life" (15). Circus becomes the "place for working out a new mode of interrelationship between individuals," a mode of interrelationship rooted in the deferral of common cultural signifiers such as fixed meanings, inscrutable classifications, established linguistic patterns, and conventional routines and behaviors (123).

In correlating Bakhtin's concept of carnival with Dickens's representation of circus, I suggest how Dickens may have not only anticipated the social problems accompanying the growth of technologies of production but may have also prescribed culture through the fictional rendering of circus as a remedy for attenuating technical progress by enabling culture to transcend the limits of technical boundaries. By connecting Dickens's circus to Bakhtin's carnival, one may locate the spirit of culture within the socioeconomic metaphors of *Hard Times*, metaphors that represent the grotesque transformation of body, mind, and language and their resistance to physical and ideological containment within the Victorian public sphere.

Although the circus has origins in many civilizations, its connection to Roman culture bears relevance to my work. Since early modern times, *circus* and *carnival* have become synonymous and interchangeable terms. However obvious their relation may be, there are historic and cultural differences between them that should be acknowledged. Some etymologists and orthographists derive the term *carnival* from the combination of the Latin words *caro* (i.e., meat) and *vale* (i.e., good-bye). Compounding the meanings of the terms *good-bye* and *meat*, or "good-bye to meat," supports conceptions of carnival as a period of preparation for Lent. However, the Latin *carrus navalis*, or "ship of wheels," enjoys a closer thematic resonance to contemporary conceptions of the circus as an impermanent and nomadic show, similar to Sleary's traveling circus.

The circus of Coketown is somewhat unlike the ancient circus or carnival from which it derives its name. *Circus*, a Latin word derived from the Greek word *kirkos*, meaning "circle or ring," was composed of spectacles of violent entertainment. Throughout antiquity, carnivals involved feats of daring and strength such as chariot races, staged battles with exotic animals, and gladiatorial combat. Violence, brutality, and bloodshed were common spectacles at ancient circuses or carnivals. State sanctioning of these events enabled common citizens to engage in moments of radical behavior that satisfied, in a limited sense, desire for subjective individualism. By allocating physical space for the performance of antisocial behavior, Roman authorities were able to regulate patterns of antinormative behavior. Authorizing the time and place for performances of radical behavior allowed Rome to control its appearance and duration on the cultural stage. Enabling the expression of antisocial attitudes and behaviors was a gesture toward realizing individualism. However, the gesture is nothing more than a false representation of individualism, which obscures the ways in which commonality and homogeneity are achieved.

In ancient times, carnival was the period of celebration in Roman Catholic communities, preceding fasting and tithing of Lent. Modes of parody and perversion such as masquerades, parades, and pageants emphasized carnival's connection to pre-Christian pagan rites of rebirth and renewal, rites which anticipated seasonal changes such as the coming of spring. Older circuses like Circus Flaminius, one of the first Roman circuses, had no permanent physical structure or arranged seating. This type of nomadic show begged for an audience that would anticipate its arrival as a cultural phenomenon. By creating more permanent physical structures like the Coliseum, among other arenas and amphitheaters, Roman authorities were able to regulate the appearance and timing of public festivities. Nomadic circuses became recurring patterns of public activities authorized by the dominant culture. Monstrous edifices such as the Coliseum, which contained shows like Circus Maximus, became an imposing physical and social structure within the ancient geopolitical landscape of Rome that transformed the circus from cultural phenomenon to socially accepted event.

Roman circuses evolved into institutionalized versions of carnival, social structures that were sanctioned by authorities and anticipated within communities as acceptable modes of cultural expression. Roman circuses were a popular form of entertainment and escapism. In the second century AD, the Roman poet Juvenal wrote in his *Satires*, "The people that once bestowed commands, consulships, legions, and all else, now concerns itself no more, and longs eagerly for just two things—bread and circuses."[10] The circus became a means for satisfying the demand for cultural and heritage activities across all classes and genders of Roman society. It also enabled the possibility of subverting conventional norms and values through costuming, gender bending, role-playing, and the inversion of social hierarchies. It was through this process of inversion that individuals gleaned moments of differentiation within their human condition.

At the peak of Roman power and influence during the fourteenth and fifteenth centuries, celebrations of carnival such as Bacchanalia, Saturnalia, and Lupercalia enabled moments of temporary freedom from social restrictions. The sanctioning of licentious behavior augmented levels of social unrest and civil conflict. Public festivities of carnival were regarded by the Roman Catholic Church as untenable modes of resistance to secular authority, so much so that the pagan celebrations were restricted by the church in the late sixteenth century. Forms of corporal punishment such as flogging and pillory were displayed in commercial areas not only to decrease criminal activity but also to remind carnival participants of their obligation to adhere to social restrictions on behavior.

Through perpetual movement, early modern circuses hoped to avoid state control and punishment, becoming transitory and nomadic modes of cultural expression. By locating themselves on the periphery of society, circuses limited the possibility of intervention within the culture. Dickens repeats this idea by locating Sleary's circus along the fringes of Coketown. However, its presence is anticipated by characters such as Tom and Louisa, who sneak into the circus not only to satisfy their curiosities but to engage also an alternate reality. As Linda Hutcheon suggests, "In discussing the particular case of the medieval carnival, Bahktin seems to have uncovered [an] underlying principle of all parodic discourse: the paradox of its authorized transgression of norms" (74). Carnival "posits, as a prerequisite to its very existence, a certain aesthetic institutionalization which entails the acknowledgment of recognizable, stable forms and conventions" (74–75).

Carnival tests the limits of social and ideological boundaries. It enables individuals to transcend prohibitions imposed by dominant culture. By allowing for the transgression of boundaries, carnival provides a glimpse of differentiation from the norm, from experience of other as representation of human condition. Yet, it provides the knowledge that one may return to the safety and security of the familiar, a position of comfort located within the confines of Sleary's circus.

The fictional world of Coketown mirrors that described by Peter Stallybrass and Allon White in *The Politics and Poetics of Transgression*.[11] Their descriptions suggest how Victorian "bourgeois fears and anxieties" about working-class disruptions were projected onto the grotesque body of the modern, industrial city "with all its threatening impurity and [restrictive] borders" (Messenger; Dickens 1.1.8, 1.3.22–23). Its vile landscape informed bourgeois conceptions of the impurity of its inhabitants and also represented their physical confinement within institutional systems of control such as the factory. Spatial containment within the factory system limits the movement of workers. However, the formation of a new technical and factual language enables the policing of ideas and the separation of mind and body through industrialization of labor. Within the factories of Coketown, workers lose the ability to make thoughtful decisions. Rather, workers respond to the consequences of their human condition and to the factual language that informs their dreary existence.

The embodiment of labor requires consciousness that is made recognizable and expressible through human presence. This physical presence as a signifier of consciousness becomes the commodity of human labor. As workers perform specific tasks, they are routinized to the demands of labor and conscious of their individual contributions to the overall processes of production. The value of human agency is accentuated as workers assume responsibilities tailored to their skills and attributes. However, technical intervention into sensory experience severed the connection between presence and consciousness, and Victorian perceptions of the external world were reconceptualized in terms of an emergent technical language. Bounderby's description of his mill workers as "Hands," for example, is a playful metonym to suggest their inseparable unity as individual components of a greater mechanical whole. On a deeper level, it represents the five senses used to inform perceptual imagination.

By way of the hand, we are able to touch objects that are seen. Tactile experiences alter our relation to objects and necessitate the decoding of experiences within our perceptual imagination. The reconstruction of tactile experiences takes place when perceptions are disembodied and recoded in memory as representations. But in some ways, the structural process of representation is more complex. The characters within Coketown exist in a double reality that lacks a double consciousness. On the one hand, characters such as Sissy Jupe profess their desire for cultural renewal and the pluralization of consciousness. On the other hand, characters such as Bounderby and James Harthouse, a disciple of Gradgrind's rigid philosophy who comes to Coketown from London in order to become involved in local politics as a diversion from his boring existence, perceive of themselves in relation to the hierarchical system of a singular, technical consciousness which focuses on "trust" and "political economy" (1.8.40). This doubleness is emphasized throughout the novel, particularly when characters attempt to navigate the boundaries between the circus and the factory system as antagonistic social formations.

The workers of Coketown are beholden to Bounderby and other mill owners for their sustenance and the maintenance of their existence. They know they are components of the factory system, but they are also aware that they might be disposed of or replaced by technologies of production. Bounderby's classification of workers as "Hands" is a reminder of the contribution of manual labor to production. It also alludes to the ways in which the human body has been disconnected from production, and disembodied through the automation of technical processes.

The representation of workers as instruments of technical processes suggests the power of perceptual disembodiment. Each finger's movements are synchronized to the movements of the adjoining fingers, and each fulfills the purposes of the entire hand. Each worker's functions are symmetrically aligned to advance routines of production, routines that become rhythmic in their efficiency and expediency. Their language is communicated through technical principles devoid of critical inquiry or spontaneous emotion. Their purpose is demonstrated by their suboptimization to technologies of production. Their sense of community is acquired through routinization and habituation of labor. When the patterns are disrupted, as in the case of Stephen Blackpool, whose loyalty and diligence are tested by his unwillingness to choose between the workers and Bounderby, identity is negated and the physical body is expelled and purged from the system of Coketown.

In contrast, Sleary's circus members are an organic counterculture that evolves in response to perceptions of social and industrial reengineering of the eighteenth and nineteenth centuries. Dehumanizing patterns of labor and social behavior allowed for the objectification and fetishization of culture during the Victorian era. However, it is the representation of Sleary's circus as counterculture which makes critique available for intervention into systems of technical rationality, systems advocated not only by mill owners like Bounderby but also by public educators like Gradgrind.

It is through the connection between factory and school and between systems of labor and systems of learning that we see the ways in which metonymic signs are used to conceptualize how members of social groupings experience liminality. In its opening chapter, the novel satirizes early Victorian conceptions of education and the manner by which learning experiences were systematized to support the memorization of facts. Gradgrind, master of the only school in Coketown, emphasizes the value of the factual language used to educate students:

> Now, what I want is, Facts. Teach these boys and girls nothing but Facts. Facts alone are wanted in life. Plant nothing else, and root out everything else. You can only form the minds of reasoning animals upon Facts: nothing else will ever be of any service to them. This is the principle on which I bring up my own children, and this is the principle on which I bring up these children. Stick to Facts, sir! (1.1.3)

When Gradgrind suggests "nothing else will ever be of service" to his students but factual knowledge, he demonstrates the limits of his rationalist philosophy, a philosophy rooted in the theories of nineteenth-century Utilitarian thinkers such as Jeremy Bentham or James Mill. Both theorists emphasized the value of empiricism and rationalism as means for regulating and predicting patterns of human behavior. Gradgrind brings this philosophy not only to teaching but also to child-rearing.

His children, Tom, Louisa, Jane, Adam Smith, and Malthus, demonstrate the problems of Gradgrind's strict adherence to a utilitarian philosophy. Their lack of emotional connection to their father is the result of his depriving them of fantasy and feeling. As children grow, the opportunity to engage in play, to perform roles, to assume other voices, and to experience alteration of reality broadens their understanding of their relation to the world around them. Gradgrind's philosophy restricts experiences of creativity, individualism, and differentiation and limits the potential for questioning the authoritative voice of conventional wisdom.

Gradgrind represents himself as "a man of facts and calculations. A man who proceeds upon the principle that two and two are four, and nothing over, and who is not to be talked into allowing for anything over" or beyond the factual designations he employs to represent himself and other characters (1.2.6). As Gradgrind administers learning to his students, he eschews and restricts their subjective identities by assigning them numerical designations.

By addressing his students as numbers rather than as individuals, Gradgrind lessens their significance and suggests their minor role as a part of a larger collective. When Gradgrind calls out students by their assigned numbers, he represents the absence of subjective identity. The students' acquiescence to their new designations establishes class objectives to be set forth in factual terms: "In such terms, no doubt, substituting the words 'boys and girls,' for … [numbers,] Thomas Gradgrind now presented Thomas Gradgrind to the little pitchers before him who were to be filled so full of facts" (1.2.6). However, by rejecting their numerical identification (e.g., "Girl number twenty … Don't call yourself Sissy. Call yourself Cecilia" [1.2.6]), students may interrupt the flow of facts and the intrastudent regulation of thoughts and behaviors as the essential objective of class lectures. Sissy challenges the ways in which technical language distorts or eschews subjective identity and devalues the uniqueness of human personality.

During a class discussion, Gradgrind asks Sissy to provide him with a "definition of a horse" (1.2.7); despite having watched her father ride and tend to horses as a circus performer, she is unable to provide a factual definition. Sissy imagines the aesthetic beauty of a horse and conceives of how she might wallpaper her room with pictures of horses. However, she is unable to state a factual definition because her expressiveness cannot be easily calculated or quantified. Sissy is rebuffed by her fellow student, Bitzer, who embodies Gradgrind's philosophy and repeats the factual definition he has learned in school:

> The definition of a horse that Bitzer offers relies upon the implicit claim that language can account absolutely for

> the things it names, so that, having heard the definition, Sissy is expected immediately to "know what a horse is" (book 1, chap. 2). In Gradgrindery, the assumption is that, because signs can substitute absolutely for things, they are indistinguishable from them—therefore, since horses do not walk up and down walls in reality, you should not paper walls with representations of horses and, since you don't walk over flowers in reality (an odd assumption, this, for Gradgrind), you shouldn't put representations of flowers in a carpet. Gradgrind's rage for substitution means that he can conceive easily of perfect translations of one sign into another—" 'What is called Taste, is only another name for Fact' " (book 1, chap. 2). (Connor)

Sissy is chastised by Gradgrind for using her imagination or "Fancy" to supplement "Fact," and, as he suggests, "You are to be in all things regulated and governed … by fact" (1.2.9). The conflict between Sissy's imagination and Gradgrind's factual determinacy underscores how being able to properly assign designations is an essential part of validating information. It also demonstrates the ways in which transitive grammar objectifies the individual by making names irrelevant designations. Thus, "facts" are nothing more than choices one makes of how one wishes to perceive of things. But, if one's choices are derived only from facts, how might one conceptualize that which exists beyond the limits of certainty? How might one engage in experiences of the arts, such as literature?

Sleary's philosophy of amusement and imagination is a corrective to the lifeless human condition of mid-Victorians, a condition facilitated by technologies of production and the technical language that defined them:

> "People mutht be amuthed, Thquire, thomehow," continued Sleary, rendered more pursy than ever, by so much talking; "they can't be alwayth a working, nor yet they

> can't be alwayth a learning. Make the betht of uth; not the wurtht. I've got my living out of the horthe-riding all my life, I know; but I conthider that I lay down the philothophy of the thubject when I thay to you, Thquire, make the betht of uth: not the wurtht!" (1.7.35)

The demand for amusement is suggestive of the ways in which culture may become a cathartic remedy for the human condition. This cultural deficiency is demonstrated by the working class of Coketown, who crave culture and express a disinterest in the technical prose of mathematicians such as Euclid and Edward Cocker. Their preference for literature, such as Daniel Defoe's *Robinson Crusoe* and *The Vicar of Wakefield* by Oliver Goldsmith, emphasizes their resistance to the imposition of factual language and culture:

> There was a library in Coketown, to which general access was easy. Mr Gradgrind greatly tormented his mind about what the people read in this library: a point whereon little rivers of tabular statements periodically flowed into the howling ocean of tabular statements, which no diver ever got to any depth in and came up sane. It was a disheartening circumstance, but a melancholy fact, that even these readers persisted in wondering. They wondered about human nature, human passions, human hopes and fears, the struggles, triumphs and defeats, the cares and joys and sorrows, the lives and deaths of common men and women! They sometimes, after fifteen hours' work, sat down to read mere fables about men and women, more or less like themselves, and about children, more or less like their own. They took DeFoe to their bosoms, instead of Euclid, and seemed to be on the whole more comforted by Goldsmith than by Cocker. Mr Gradgrind was for ever working, in print and out of print, at this eccentric sum, and he never could make out how it yielded this unaccountable product. (1.8.42)

If society has failed in its attempts to delegitimize carnivalesque and must now allow it to occupy its own physical and ideological plane, must it also allocate room for antagonistic and controversial types of literature as well?

As noted by Victorian scholar Nancy Armstrong, the reading habits of the Victorian working class demonstrated an antagonistic response to the substitution of statistics and facts for imagination and creativity:

> If the reading of fiction came to play an indispensable role in directing desire at certain objects in the world, it was not because such narratives as *Robinson Crusoe* administered a particularly useful dose of didactism … [rather, it suggests the] possibility that moral hegemony triumphed in nineteenth century England largely through consent rather than coercion; it was precisely because they were leisure-time reading that such books as *Robinson Crusoe* were important to the political struggle between the ruling classes and the laboring poor. (16)

Victorian issues about education and literacy among the working class are emphasized by Gradgrind's reaction to the reading preferences of the working class who patronize Coketown's library. Rather than regarding their enthusiasm for any type of literature as beneficial to improving education through literacy, Gradgrind expresses paranoia toward literary realism and regards literature as a device of repression which serves the interests of dominant culture. As Catherine Belsey explains,

> [B]y … taking for granted what is usually taken for granted, by replicating … reality without questioning it, [one] collaborates in the work of ideology: the naturalization of the social, the eternization of the historically derived, the installation of but one version of reality as reality itself.[12] (82)

Does the working class' choice of literature promote antagonism toward social norms or prevent the possibility of resistance? And if certain types of literature do promote resistance to society, how might education and literacy remind us of the ways in which culture frees us from a lifeless and mechanized existence? Michel Foucault suggests the novel is "part of that great system of constraint by which [Western culture] compelled the everyday to bring itself into discourse" ("Infamous Men" 91), and it is through even the most mundane forms of discourse that the possibilities for altering one's human condition are made obvious and achievable.

Coketown's reading habits suggest not only a rejection of statistics and geometry but also an enthusiasm for experiences of wonder and imagination, attributes found common in not only Sissy but in Gradgrind's daughter Louisa as well. The conflict between fictional literature and factual data is further suggested when Louisa observes the flickering shadows cast on her bedroom wall by her fireplace and begins to weave together the tattered fabric of imagination to narrate an entertaining story for Tom:

> "Except that it is a fire," said Tom, "it looks to me as stupid and blank as everything else looks. What do you see in it? Not a circus?"
>
> "I don't see anything in it, Tom, particularly. But since I have been looking at it, I have been wondering about you and me, grown up."
>
> "Wondering again!" said Tom.
>
> "I have such unmanageable thoughts," returned his sister, "that they will wonder." (1.8.45)

Although he remains attentive to her tale, Louisa's brother is neither able to envision nor to embrace her story because his perceptions are limited to fact-based reality which eschews the possibility of wonder and imagination. Coketown is the "triumph of fact" over fancy and the vindication of the "severely workful" over the perpetual dreamer (1.5.20). Tom's strict upbringing, which stressed the value of rational self-interest, prepares him for the routines he will perform as apprentice at Bounderby's bank. His desire to acquire wealth as a means of escaping the limitations of his human condition is evidenced by his habitual gambling and his indifference toward financial loss, particularly that which is experienced by other characters. For Tom, monetary gains are nothing more than commodities of exchange. Their limitation is in his ability to possess and retain them. Unlike Tom, Louisa is able to engage in a process of storytelling, through which she renders fictional realities as alternate possibilities of existence.

In his study of the effects of Sunday schools on nineteenth-century working-class education, Thomas Walter Lacqueur determined, as Armstrong interprets, "literacy did not simply indoctrinate the poor in the values and practices that would make them fit to inhabit an industrial world" (Armstrong 17). "If education helped to produce a more tractable working class," as Armstrong correctly concludes, then "working-class radicalism was predicated on literacy too—that is, on political pamphlets, on alternative programs for education, and even on a literature that spoke to their needs and desires rather than to those of their employers" (17). The workers' literary preferences demonstrate the deferral of reality through fiction, and

> [i]t is appropriate that the story that Sissy remembers telling her father should be that of Scheherazade, in which a princess staves off her execution by telling a succession of stories; it is a narrative which is actually about the deferment of reality by an excess of narrative. But even realistic fiction, "about men and women, more or less like

> themselves, and about children, more or less like their own" is regarded by the Coketown workers as a sort of relaxing addition to their lives, rather than an inert reflection of them, and Gradgrind is perplexed "at this eccentric sum, and he never could make out how it yielded this unaccountable product" (book 1, chap. 8). (Connor)

Gradgrind's inability to acknowledge the value of literary fiction is a reminder of the intractableness of his system of logic, a way of thinking that obscures morality in favor of objectivity. However, Gradgrind's reasoning is placed in jeopardy while visiting Sissy's home, when his rigid representation of Sissy proves false and untenable.

When Gradgrind pays a visit to Sissy's home, he ponders how she has been unable to adapt to or to inhabit his industrial world and, thus, chooses to reject her as an unmanageable thought within his own factual discourse. However, when he observes in Sissy's home life the absence of parental supervision, the designations he applies to Sissy are contradicted as absolute facts and are made irrelevant. Upon learning Sissy's father has left her, for the first time Gradgrind experiences a momentary lapse in factual certainty. The logic by which he judges Sissy is made erroneous, and his own ways of thinking are placed in question.

Sissy and her father, a clown in Sleary's circus, live at the Pegasus's Arms, which is a public house patronized by members of the circus. The interior of the public house is quite dramatic and expressive. Its interior is adorned with theatrical symbols that represent circus culture, such as the Pegasus, "with real gauze let in for his wings, golden stars stuck on all over him, and his ethereal harness made of red" (1.6.25). The symbols stimulate revelry and merrymaking and also remind the occupants of the public house of their own culture and performance traditions. However, the joviality of the public house conceals

Sissy's abandonment by her father, who leaves her in the hopes that she may find a better life without him.

Gradgrind and Bounderby visit Sissy at the Pegasus's Arms with the intention of dismissing her from school, but Gradgrind cannot bear to abandon her as her father has already done. As members of Sleary's company gather together to observe Sissy's interactions with Bounderby and Gradgrind, the members' commitment to their community and to the social connections that unify them are demonstrated through the art of performance:

> The father of one of the families was in the habit of balancing the father of another of the families on the top of a great pole; the father of a third family often made a pyramid of both those fathers, with Master Kidderminster for the apex, and himself for the base; all the fathers could dance upon rolling casks, stand upon bottles, catch knives and balls, twirl hand-basins, ride upon anything, jump over everything, and stick at nothing. All the mothers could (and did) dance, upon the slack wire and the tight rope, and perform rapid acts on bare-backed steeds; none of them were at all particular in respect of showing their legs; and one of them, alone in a Greek chariot, drove six in hand into every town they came to. They all assumed to be mighty rakish and knowing, they were not very tidy in their private dresses, they were not at all orderly in their domestic arrangements, and the combined literature of the whole company would have produced but a poor letter on any subject. Yet there was a remarkable gentleness and childishness about these people, a special inaptitude for any kind of sharp practice, and an untiring readiness to help and pity one another, deserving often of as much respect, and always of as much generous construction, as the every-day virtues of any class of people in the world. (1.6.30–31)

As Sissy mourns the loss of her father, Bounderby grows impatient with her and tells Gradgrind to leave her at the public house. Bounderby's willingness to abandon Sissy is received harshly by the circus members, who

> cared so little for plain Fact, these people, and were in that advanced state of degeneracy on the subject, that instead of being impressed by the speaker's strong common sense, they took it in extraordinary dudgeon. The men muttered "Shame!" and the women "Brute!" and Sleary, in some haste, communicated the following hint, apart to Mr Bounderby. (1.6.32)

Gradgrind resists Bounderby's suggestions and offers Sissy the opportunity to live with him and his family, provided she never again visit the circus or interact with any of its members. Sissy, at the behest of Sleary's daughter Josephine, accepts Gradgrind's offer. Though Gradgrind welcomes Sissy into his home, his motives are not selfless. It is Gradgrind's hope that Sissy may become "an example to Louisa, of what this pursuit which has been the subject of a vulgar curiosity, leads to and ends in" (1.6.30).

For Gradgrind, Sissy becomes anecdotal evidence as a representation of human failing and of striving for a flawed life within the circus. Gradgrind attempts to reeducate and to convert Sissy to his way of thinking. However, "[l]ife by its nature is dialogic. To live means to participate in dialogue: to ask questions" and to explore the limitless possibility of alternative choices that confront us (Morson and Emerson 60). Utilitarian philosophy presumes that humans have reached the limits of all that is conceivable, and thus, utopianism is achieved.

Theoretically speaking, utopianism's one essential quality is its unattainability. Utopian society is not ever achievable. If it were, it would necessitate our admitting that society or, for that matter, humanity has reached its zenith. Consider how listless society would become if there were nothing more to be known, nothing more to be learned, nothing more to be conceptualized or experienced. History would come to an end, since history is the measure of human progress and all progress would become irrelevant in perpetual utopia. Culture and language would lose their ability to reconstitute themselves and to reinvigorate the human spirit. Thus, conceptions of the

> utilitarian utopianism of Comte and St. Simon, in which all culture progresses reliably through the stages of supernatural and metaphysical development finally to come to poise in the scientific humanitarianism of positivism; and the schemes of various British Victorian constructors of cultural theories, including Carlyle, Mill, and Arnold [See Carroll, Evolution and Literary Theory 21–24, 184–91, 200–201.] (Darwin)

All human motivations would be adjusted to assure the preservation of the greater good through constancy.

When Gradgrind calls his students "reasoning animals," he suggests how their minds may be cultivated as the fertile soil in which factual systems of knowledge may be planted to yield a rich harvest of technically evolved humans, essential components of the factory system. The repetitive exercises conducted by Gradgrind's instructor, Mr. Choakumchild, emphasize learning through routinization and habit. The curriculum at Gradgrind's school prepares students for the rigors of factory labor by teaching them to devalue their imagination and to accept the obligations of the mechanized roles society has already assigned to them. In effect, Gradgrind's students will mature and will repopulate the ranks of the working class. As each year passes, students progress through their education in preparation for the roles they will assume as factory workers, roles performed by their parents. "[The] plain, bare, monotonous vault of a schoolroom" replicates the dreary environment of Bounderby's mill and conditions the children to acclimate themselves to the subhuman public spaces they will come to occupy within industrialized Coketown (1.1.5).

Unlike Bounderby's mill workers, Sleary's circus is a heterogeneous grouping with an uncommon language, conflicting purposes, and senses of communitas. Their apparent lack of commonality parallels the ways in which manual laborers experience separation from technical processes as a loss of socialization, a loss of language, and a loss of belonging within the factory system. Yet, unlike the workers of Coketown, whose common experiences within the factory system sever and isolate them from humanity, the circus members' differences keep them together and enable them to project alternate

perceptions of reality, perceptions that do not "acquiesce to the legitimacy of the present system" (Gardiner 32).

The circus enables the location of physical spaces of transgression through which one may find opportunities to participate or to observe role-playing and the alteration of identity, much like that found in Bakhtin's carnival. Within the circus, parodies of normative life are rendered through acts of performance, parodies that suggest deficiencies within dominant conceptions of reality. However, rather than considering the circus as a mode for expressing utopian ideals, I wish to align it more closely with Foucault's conception of *heterotopia* (in "Of Other Spaces").

Foucault defined heterotopia as a "counter-site" to "a kind of effectively enacted utopia in which the real sites, all the other real sites that can be found with the culture, are simultaneously represented, contested, and inverted" ("Of Other Spaces" 24). For Foucault, utopia is an ideological reality that is left unfulfilled and made not achievable by its absence of location. Utopias are ideas that lack a physical representation and thus eschew signification, whereas heterotopia is the ideal unreality, a physical and ideological point of intervention within the dominant reality. Sleary's circus is sanctioned and authorized by the dominant culture and locates itself within the same ideological system its signification should refute.

The circus becomes a "widely available commodity divorced from didacticism, subject to its audience's critical scrutiny, and vulnerable to the disciplinary forces of an increasingly regulated society" (Schaeffer 342). However, its existence depends on its affirmation of the dominant culture it seeks to resist. Yet, both the circus and dominant culture of Coketown retain their imminent totalities. Both exist in a perpetual state of becoming that nullifies the possibility of ever realizing utopian dreams.

Bounderby's workers experience communitas, language, and belonging as nothing more than a simulacrum, an illusion of socialization legitimated by epistemes and new a priori conditions which abstract and reduce human agency within the Victorian technical sphere. The "Hands" and Sleary's circus become metonymical signs of themselves, because they must be perceivable and relatable to the members who compose them. The aesthetic of the circus, for example, must be reducible to abstract qualities that represent its human parts. But, unlike representations of the factory, which devalue workers as replaceable components, the differentiation of circus members assures their irreplaceableness and revalues their performance of unique roles within the circus.

The circus, like Bakhtin's carnival, "makes it possible to extend the narrow sense of life" beyond systems and ideologies (*Rabelais* 177). The circus locates itself within the periphery of Coketown, where certainty gives way to inexactitude, where permanence gives way to mutability, and where normative order yields to formations of anarchy, disorder, and chaos. It is from within this range of instabilities that stability is actually achieved through differentiation. The circus exists, therefore, as a reminder of "utopian potential … [which] lies not in the reclaiming of our 'nature', not in adopting a fictitious position 'outside', but in our ability to change within the structures attempting to define us" (Bammer 28). As Foucault observes,

> The critical ontology of ourselves has to be considered as an attitude, an ethos, a philosophical life in which the critique of what we are is at one and the same time the historical analysis of the limits that are imposed on us and an experiment with the possibility of going beyond them. ("What is Enlightenment?" 89)

When you purge what is unique in an individual, you also purge the desire for dissent. And, when you purge individuals of the desire for dissent, you purge them also of the desire to differentiate. The circus, therefore, demonstrates how individuals may be true to themselves and true to the collective.

The circus enables the experimentation of identity by "extend[ing] our participation" in the dominant culture through parody, perversion, and violation of rules, norms, and values (Foucault, *Language* 230). "People who in life are separated by impenetrable hierarchical barriers enter into free and familiar contact on the carnival square" (Bakhtin 123). During carnival, traditional social hierarchies are reconfigured to support the performative act of role-playing and the mutual exchange of identities that "[lie] beyond existing social forms" (280). At carnival, representations of gender and class are destabilized, corrupted, and resignified to suit new modes of social interaction, and this process of resignification is demonstrated through performative acts, such as when Sleary's circus performers alter their physical appearance to signify new identities.

Individuals engage in play by altering their physical appearance and by assuming new roles through differentiation of identity. As individuals reconstitute their identities, they reclaim the right to differentiate themselves, "the right to be 'other' in this world" (Bakhtin 159). The reconstruction of social reality within Sleary's circus enables "the promise of a renewal of humankind on a more egalitarian and radically [new] basis" (Gardiner 30). Circus, like carnival, is "the place for working out a new mode of interrelationship between individuals" (Bakhtin 123), and the reconfiguring of social relationships is expressed when Tom flees to the circus.

At Sissy's behest, Tom travels to the circus in order to escape punishment for not only robbing Bounderby's bank but also for implicating Stephen Blackpool in the theft. Tom changes his physical appearance, performs staged routines, is discovered, and is eventually captured and interrogated by Bitzer. With the assistance of Sleary, his performers, and his trained circus animals, who themselves perform tricks to distract the authorities, Tom flees his pursuers to the port of Liverpool. Despite his circumstances, Tom is inevitably able to reclaim his freedom and escape to a new life with the help of Sissy, Gradgrind, and Louisa, who plan his passage aboard a ship that will take him beyond the reach of guilt and accountability in Coketown.

The circus performers help Tom despite his presumed guilt. Theirs is a compromise between morality and loyalty, between "rule of law and order of custom."[13] Dickens positions the circus as a sanctuary insulated from the encroachment of dominant culture. However, its location within the fringe areas of Coketown (and towns like it, for that matter) reminds us of its need for continuous legitimation and authorization within the dominant culture. As a fragment of hope within the dominant culture, the circus may be but a reminder of primitive society, in which customs, traditions, and rituals were performed as a means for reaffirming social expectations and moral obligations:

> Efforts to legislate conscience by an external political power are the antithesis of custom: customary behavior comprises precisely those aspects of social behavior which are traditional, moral and religious—in short, conventional and nonlegal. Put another way, custom is social morality. The relation between custom and law is basically one of contradiction, not continuity. (Diamond 257)

The circus enables the deferral of absolute meaning and intractable signification. It is, in many ways, the acknowledgment of alternative possibility, the preference of social morality to factual law, and the human desire for resistance to cultural permanence.

The circus is an anomaly located within Coketown's rigid systems of rationality. And, as an anomaly, it is a reminder of the inherent flaws within these systems. Its own authorization to exist within the periphery of Coketown is a strategy employed by authoritative systems for avoiding representations of their own apparent limitations and contradictions. By allocating social spaces within these systems for institutionalizing anomalies like Sleary's circus, these systems become aware of their own structural deficiencies and falsely represent those deficiencies in order to prevent full ideological disclosure, a process that begins through the recognition of anomalous cultural formations that cannot ever be fully purged from within the dominant culture.

CHAPTER 3

CROWDS, MOBS, AND SPATIAL SEPARATIONS IN BRONTË'S *SHIRLEY*

> Zarathustra grasped the tree beside which the young man was sitting and spoke with him thus: If I wanted to shake this tree with my bare hands I should be unable to do it. But the wind, which we cannot see, torments and bends it where it wishes. It is invisible hands that torment and bend us the worst. Now it is with men as it is with this tree.
>
> —Frederich Nietzsche

Shirley takes place between 1811 and 1812 in rural Yorkshire, and the 1811 Luddite riot against the suboptimization of human labor serves as the historical backdrop for Charlotte Brontë's novel. The radical sentiments of the Luddites are continually reinvigorated as Brontë's fictional crowd becomes a literal character, lurking always on the periphery of the novel. Its presence is made possible by the ways in which characters anticipate its presence as a physical reality, even in its absence. As Brontë's characters negotiate the realities of crowd formations, they also make decisions about their degree of public involvement. In so doing, they acknowledge the inevitability of crowd participation and suggest the development of a public sphere of engagement in which individuals are no longer able to express singularity but rather embrace the plurality of movement among crowds, and in so doing, "no longer stand alone where all others went in pairs or parties" (3.6.312). The relationships and subsequent love triangle that develops between the characters of Caroline Helstone, Shirley Keeldar, and Robert Moore serves as a plot device for advancing the narrative, but it also provides examples of how crowd participation enables them to cultivate their relationships by managing with whom they are associated. It is through associative symbols and meanings that crowds are better able to organize, to mobilize, and to radicalize human behavior toward a collective interest. Brontë historicizes this process within the context of the Luddite movements of the early nineteenth century.

The Luddites were named for a supposed Leicester stockinger's apprentice, Ned Ludham, who destroyed his master's stocking frame in order to resist the mechanization of labor. In February of 1811, the Luddite movement began in the Midlands between Nottingham, Leicester, and Derby. Outbreaks of violence continued to spread throughout mills and factories as Luddites galvanized support from local communities. In January of 1812, northern cities like Lancashire, Cheshire, and Yorkshire experienced violence and machine-breaking as the reaction of the working class to new technologies of production, technologies which altered workers' contribution to and control of labor.[14] Frank Ongley Durvall suggests in his book *Popular Disturbances and Public Order in Regency England*,

> Most people in 1811 and 1812 found it difficult to appreciate the value of new machinery economizing labor at a time when goods were a glut upon the market and when there was, in any case, a surplus of labor available. (62)

Victorian laborers' struggle to regulate technological intervention within the workplace was equaled by their desire to revalue human capital within the emergent factory system.

The formation of crowds was a direct response to the organization and mechanization of labor. By utilizing the power of the crowd to speak to Parliament and to entrepreneurs of industry, Victorian workers communicated from the crowd, through the crowd, and beyond the crowd in order to transcend it, to regulate its movements, and to utilize its physical energies to achieve working-class demands. This chapter considers how speaking from the crowd may empower subjective interiority while heightening individual desire to distance oneself from the crowd's voice, which remains inarticulate and illegitimate without the collectivization of individual speech and action. It is from within hidden spaces of crowds that individuals are able to secure subdivision, extension, and propagation of self from others, while simultaneously experiencing abstraction, isolation, and reduction from the same signifiers.

One of the more interesting aspects of Brontë's industrial novel is its depiction of working-class crowds as both physical entities and metonymical symbols. Crowds have often been meagerly represented as peripheral groups, while more well-rounded characters fill the foreground of Victorian novels. Characterizations of Victorian crowds have either celebrated or condemned modes of group socialization commonly practiced during the Industrial Revolution. Binary formulations such as civilized or barbaric, intelligent or mindless, and passive or aggressive have been used by Victorian novelists like Brontë to differentiate crowd typologies. These formulations, however, do not adequately represent the complex arrangement of individual identities within crowds, identities which mobilize group formations and affect actual and contingent behavior.

The doubling of oppositions allows readers to distinguish conceptions of "the crowd" and "the mob" as both physical presences and symbols of signification within the novel. Characters such as Shirley Keeldar and Robert Moore employ linguistic techniques to navigate crowds, to stabilize their identities, and to establish the terms of their relationships by aggressively asserting control over conversational space, while other characters, such as William Farren and Caroline Helstone, retreat within their safety zones and become reduced and abstracted through language. Throughout the course of their conversations, contexts that would reveal intended meanings are obscured not by inexactitude but by the disharmony of meanings and objectives. For example, Shirley is quite sure of what she wishes to say and how to express it but exercises restraint and allows the opportunity for linguistic power exchanges to take place. Robert, on the contrary, struggles to withhold speech and suffers from unwillingness to partake in any exchange of power. Without linguistic power exchanges, speech becomes meaningless, social bonds weaken, and nothing may be accomplished. In Brontë's novel, characters' participation within crowd formations shows how linguistic exchanges take place and determines whether subjective identity is achieved or subsumed and whether power is reciprocally transferred through the singularization or pluralization of language.

Throughout her novel, Brontë employs structures of language such as semantic patterns, silence, misinterpretations, questions, exclamations, pauses, and non sequiturs as a backdrop for examining how characters achieve movement between spatial boundaries of dialogue. Unlike their individual components, each type of crowd maintains a unique personality that acts as its own signifier. Were it not for the aggregation of individual speech, crowds would be nothing more than a sign of themselves, a symbolic representation of voiceless and speechless masses that inhabit the periphery of industrial novels.

As the novel begins, readers are confronted by the ominous appearance of a crowd formation that will continually reshape itself to satisfy new meanings and objectives used to signify its presence and potential power within the novel. Mike Hartley, a drunkard, member of the lower class, "Antinomian" (1.1.17) ("crazed ... or crafty—or, perhaps, a little of both" (1.1.16), glimpses the novel's first crowd formation, a group of Luddites drilling on the moors. When the Reverend Mr. Helstone, Caroline's uncle, relays Hartley's observations of the Luddites drilling on the moors in preparation for their revolt, Hartley's descriptions echo a "distant sort of 'beyond' for the crowd to inhabit" (Plotz 158):

> According to his account, Mike was busy hedging rather late in the afternoon; but before dark, when he heard what he thought was a band at a distance—bugles, fifes, and the sound of a trumpet; it came from the forest, and he wondered that there should be music there. He looked up. All amongst the trees he saw moving objects, red, like poppies, or white, like may blossom. The wood was full of them; they poured out and filled the park. He then perceived they were soldiers—thousands and tens of thousands; but they made no more noise than a swarm of midges on a summer evening. They formed in order, he affirmed, and marched, regiment after regiment, across the park. He followed them to Nunnely Common;

> the music still played soft and distant. On the common be watched them go through a number of evolutions. A man clothed in scarlet stood in the centre and directed them. They extended, he declared, over fifty acres. They were in sight half an hour; then they marched away quite silently. The whole time he heard neither voice nor tread—nothing but the faint music playing a solemn march… . They seemed passing Fieldhead, when a column of smoke, such as might be vomited by a park of artillery, spread noiseless over the fields, the road, the common, and roiled, he said, blue and dim, to his very feet. As it cleared away he looked again for the soldiers, but they were vanished; he saw them no more. Mike, like a wise Daniel as he is, not only rehearsed the vision but gave the interpretation thereof. It signifies, he intimated, bloodshed and civil conflict. (1.1.16–17)

The phantasmagorical crowd Hartley envisages on the moors is later transformed into the Luddite mob, a physical entity, a "composite character … with no head but many bodies" (Plotz 180). Brontë's novel places the crowds just beyond the range of definite perception, lurking ominously in the periphery of the narrative, and as literary theorist John Plotz suggests, "[I]f they signify something other than the normal agricultural flurries far from the centers of power, it is not immediately evident what [that may be]" (158). However, it should be noted that the dream-like image of the crowd is itself a sign that awaits signification through the crowd's transformation to physical presence. The dream-like vision of the crowd formation gathering on the moors may also cue readers to the many mutations of crowds within the novel.

Brontë's depiction of the Luddite mob through fantastical imagery satisfies Victorian fears and anxieties over the power of crowds as physical formations by yielding to stereotypes of crowds that oversimplify their behavior and reactive tendencies.

Brontë's descriptions of crowds change from dream-like fantasies of marching rows of fairies, poppies, and midges to physical swarms of frolicking children and, later on, to riotous mobs of workers. In so doing, Brontë alters the ways in which Victorian crowds were to be conceived and represented and also suggests new possibilities for linguistic power exchanges that are first conceptualized as dream-like states of accommodation, conciliation, and harmony but are later realized through volatile stages of social interaction.[15]

Common descriptions like "obedient crowd" or "mindless mob" deemphasize the multitexture of individual personalities that compose Victorian social groupings. By avoiding the rhetoric of overused stereotypes, one sees how the Victorian individual might be recognized as more than a mere possession of the crowd. Indeed, Victorian individuals exerted a causal influence over crowd behavior by revealing subjective interiority through radical behavior. Industrialization required Victorian individuals to commit themselves to social groups, to yield the power of choice to collective will. It is only through participation in different types of social formations that individual agency may act to transform identity by altering its representation within society. The character of Robert Moore demonstrates how individual agency may act upon society and, at times, refashion society to suit the individual's own interests:

> At the time this history commences, Robert Moore had lived but two years in the district, during which period he had at least proved himself possessed of the quality of activity. The dingy cottage was converted into a neat, tasteful residence. Of part of the rough land he had made garden-ground, which he cultivated with singular, even with Flemish, exactness and care. As to the mill, which was an old structure, and fitted up with old machinery,

> now become inefficient and out of date, he had from the first evinced the strongest contempt for all its arrangements and appointments: his aim had been to effect a *radical reform*, which he had executed as fast as his very limited capital would allow; and the narrowness of that capital, and consequent check on his progress, was a restraint which galled his spirit sorely. Moore ever wanted to push on. "Forward" was the device stamped upon his soul; but poverty curbed him. Sometimes (figuratively) he foamed at the mouth when the reins were drawn very tight. (1.2.29, emphasis added)

The reform Moore seeks to "effect" is both radical and technical. The technical reforms he uses to measure progress are counterposed by radical subjectivism, which became Victorian individuals' response to the absence of choice and the imposition of technologies of production. By exhibiting radical subjective behavior, Victorian individuals released themselves from the status quo and enabled the possibility of social alteration. It is Moore who invokes radicalism as a move "Forward" and away from the status quo. However, radical subjective movement cannot be achieved unless representations of self are differentiated from an otherness that can only be conceptualized from within the liminal spaces within crowds.

Two types of crowds populated the Victorian public sphere, the normative and the informal crowds, and their representations are found in Brontë's *Shirley.* It is important to examine the behavior of both types of crowds within the novel and to analyze and evaluate their relation to characters like Shirley Keeldar, Caroline Helstone, William Farren, and Robert Moore, characters who either perform or resist radical behavior in order to differentiate themselves from crowd representations. The normative crowd of nineteenth-century Britain was composed of members who did not share class, gender, or political affiliation but adhered to the logic of norms, rules, and values of society. Normative crowds organized for public spectacles like political debates and cultural and religious ceremonies that appealed to the masses. Although normative crowds were signified by distinct social markers, their obedience to norms and values and their fulfillment of rules and obligations demonstrate inhibition and self-control. Normative crowd members, therefore, exhibited both the desire to make choices and the willingness to inhibit themselves from rendering such choices.

In contrast, informal crowds were made up of like-minded individuals who shared social markers but departed from the ethical language used to inform reason and discourse. Informal crowds were organized by specializations of labor: for example, tradesmen and artisans. Their ethical language validated the morality of labor, and their group customs substantiated this "moral economy" by prioritizing group initiatives over social progress (Thompson 76). Their speech and actions were prefigured by a common ethical language. This language made them predisposed to psychological stages of enthusiasm and suggestibility which are found in all types of crowds. The stages condition crowd movement and organization and also forecast decision making in conflict situations. By recognizing the potential for conflict situations, individuals tend to recognize the disciplinary power of ethical speech within crowds and perceive it as a technique for administering physical or symbolic violence or for promoting cultural solidarity.

In *Culture and Anarchy*, Matthew Arnold recounts the 1866 Hyde Park riots, which postdate the Luddite riots in *Shirley*. However, Arnold's essay confirms the exercise of "strong individualism" in modern England and the "unrestrained swing of the individual's personality" as a response to social and industrial transformations precipitated half a century earlier (63). Arnold places culture in opposition to individualism or "doing as one likes" (63). From culture, Arnold derives "a practical benefit … a principle of authority; to counteract the tendency to anarchy which seems to be threatening us" (89). The anarchy Arnold presages is a reminder of the consequences of people behaving selfishly or, as he puts it,

> this and that man, and this and that body of men, all over the country … beginning to assert and put in practice an Englishman's right to do what he likes; his right to march where he likes, meet where he likes, enter where he likes, hoot as he likes, threaten as he likes, smash as he likes. (84–85)

Arnold's personal experiences of social unrest affect his resistance to radical behavior. His argument employs culture as a remedy for containing individualism, which he views as a contagion for spreading anarchy and dissolution. Culture, as Arnold suggests, exerts control over subjective interiority. Individuals experience culture as bidirectional movement, both toward self-awareness and away from society, but "above all," they experience it "as an inward operation" for self-improvement through social harmony (190). Arnold's conception of culture views the loss of radical subjective individualism as an individual's contribution to social stability and to the permanence of crowds of people "between whom the barriers have fallen" (Adorno 17). Arnold regards crowds as a "vehicle by which certain kinds of cultural meanings and conflicts are transmitted" (Munro 2). However, what might happen if and when cultural meanings are misinterpreted or their transmission is interrupted by an ethical event? In *Shirley*, the ethical event is the moment when individuals separate themselves from the crowd formation, embrace a newfound sense of individualism, and create a new language to convey their transformative experiences.

In Brontë's novel, the crowd has been portrayed, generally, by theorists like Plotz as a singular consciousness, an individual character whose actions anticipate ethical events and frequently enable those events to come into being. The ethical event is the moment when culture, language, and meaning are placed in question and might undergo alteration. Ethical events go beyond the mere measure of accountability and morality. An ethical event may also be a moment of alteration so radical and so distinct that it fails to satisfy conventional categories of meaning and requires new cultural significations. Regardless of the duration of change, the lack of signification and the absence of meaning is supplemented by an ethical language that fills the linguistic gap caused by the event. It is the ethical event that signifies the possibility of reformation of culture, language, and meaning. Because culture and language are so inseparably linked, the absence of meaning and the lack of signification are almost always accompanied by social disruption and civil unrest.

The ethical event is, therefore, the moment when the normative order of society fails to master subjectivity and is questioned, contradicted, and reconstituted along the lines of differentiation. By placing everyone within the same cultural framework, Arnold disallows the possibility of conceptual differentiation and offers us disenchanted observers who remove themselves from the dangers of crowd participation and eschew the obligations of social critique. At the midpoint of the novel, Caroline realizes such a moment of detachedness when she observes a playful gathering of one thousand Sunday-school students and is marked by fear and uneasiness. For Caroline

> knew these [children] liked her, yet she was shy even with them out of school: they were not more in awe of her than she of them: she drew near them now, rather to

> find protection in their company than to patronize them with her presence. (3.6.312)

Caroline experiences movement both toward and away from the crowd, as evidenced by her timidity and, later, by her "hurried efforts to converse" with them (3.6.312). She locates within its formation familiarity and recognition ("she could not … stand alone … so, she approached a group of her own scholars, great girls, or rather young women" [3.6.312]). She is comforted by her students' neutral response toward her ("they were not more in awe of her than she of them" [3.6.312]), a response she mimics as a sign of placement within the crowd. Her performance "*on* duty" at the communal tea urn does not mask "her evident timidity when *off*" (3.6.312; emphasis original). Caroline finds security by locating herself within a pocket of familiar faces within the crowd. She then differentiates herself from the crowd through the performance of duty. By distinguishing herself from the rest of the crowd, by performing her role, Caroline is able to acquire, if only in a momentary sense, symbolic capital to prevent the "rushing backwards and forwards" or "mutual invasion" (i.e., exploitation) of space and identity within the crowd (1.1.7). Her performance is precise and perfected to resist the possibility of nonnormative acts: "She kept too busy with her office for a while to have leisure for looking around, but the last cup being filled, she threw a restless glance over the room" while awaiting Robert's arrival (3.6.306). In that brief moment of silence, Caroline demonstrates her restlessness and her potential for radical movement, "for the impulse was very strong," yet she is "placed in her seat" and drawn back into the crowd (3.6.308). By performing her role, by fulfilling her obligations, by observing rules of etiquette, Caroline diminishes the possibility for altering her behavior and divests herself of the radical "impulse" that flows within her. In so doing, she allows her individualism to be subsumed within the crowd, for it is the radical impulse that may enable Caroline to signify a new identity through a change in behavior.

Arnold's argument fails to address the need for such moments of alteration or, for the possibility of ruptures in ethical language and reasoning, ruptures that might affect the ideological symmetry he advocates. Ruptures in ethical language lead to human exploitation, which may or may not satisfy definitions of physical violence but demonstrate the oftentimes not perceivable effects of symbolic violence. Exploitation as symbolic violence is considered in Pierre Bourdieu's works, and, as he suggests, "Gentle, hidden exploitation is the form taken by man's exploitation of man [sic] whenever overt brutal exploitation is impossible" (*Outline* 192).[16] Perhaps culture should not be conceptualized in Arnoldian terms as a panacea for social unrest. Rather, it should be reconsidered in light of Bourdieu's depiction of human exploitation through symbolic violence, a "gentle violence, which is never recognized as such, and is not so much undergone as chosen, the violence of credit, confidence, obligation, personal loyalty, hospitality, gifts, gratitude, piety" (192). The ways in which types of symbolic capital are deployed against social others suggest how unequal power exchanges signify meaning.

The amount of symbolic capital possessed by individuals determines their command of symbolic violence, which may be administered to legitimate authority and to exercise control of social others. Although Bourdieu defines three additional forms of capital (i.e., economic, cultural, and social), it is through the possession of symbolic capital that meanings are established and patterns of domination are legitimated. The product of domination is the creation of a "single linguistic community"

("On Symbolic Power" 46) that situates itself within the contexts of legal authority and government action. However, I wish to tease out a more pliant definition to such a "community," a definition derived from an ethical language that both signifies and awaits its own signification within the ethical event. By using new meanings to think of and through the event, language and thought are linked. Ethical events, therefore, "reform language, purge it of usages linked to the old society and [re]impose it in its purified form … to impose a thought [process] itself purged and purified" (31).

Ethical language becomes a mode of critical intervention, which enables individuals to recognize the possibilities of differentiation. Bourdieu's definition of a "single linguistic community" is therefore rendered not only improbable but also impossible, even as a product of domination. The very nature of language allows for ruptures and gaps, but it is the ethical event that makes these fissures more apparent and makes for greater awareness on the part of the individual. By realizing the scope of decision alternatives, however limited that scope might be, individuals begin to assess the consequences of the choices they make and begin to situate themselves or allow themselves to be situated within particular groups.

As individual behavior changes, the ethical reasons used to justify that behavior must change as well. For example, when Moore and his workers engage in a physical confrontation over the ethical reasons for employing technologies of production, they understand one another's meanings and objectives, but their agendas and their means of achieving them differ. After Moses Barraclough is apprehended by Moore for his supposed participation in the attack on Moore's mill, worker William Farren tries to convince Moore that a resolution between him and his workers is possible.

> "I've no grand words at my tongue's end, Mr. Moore, but I feel that it wad be a low principle for a reasonable man to starve to death like a dumb cratur. I willn't do't. I'm not for shedding blood: I'd neither kill a man nor hurt a man; and I'm not for pulling down mills and breaking machines—for, as ye say, that way o' going on'll niver stop invention; but I'll talk—I'll mak as big a din as ever I can. Invention may be all right, but I know it isn't right for poor folks to starve... ."
>
> "Worry the Parliament-men as much as you please," said Moore; "but to worry the mill-owners is absurd, and I for one won't stand it."
>
> "Ye're a raight hard un!" returned the workman; "Willn't ye gie us a bit o' time? Willn't ye consent to mak your changes rather more slowly? ..."
>
> "Talk to me no more about machinery. I will have my own way. I shall get new frames in to-morrow. If you broke these, I would still get more. I'll never give in." (1.18.137–138)

Although Farren's speech is clear and his meanings are apparent, as Moore's response makes obvious, his words are emboldened by the struggles of the working class and suggest their desperate need to assert control over the growth of technologies of production. These technologies rapidly diminished the value of manual labor, the only commodity the working class possesses. "[Human] [l]abor assures not only individual survival," as Hannah Arendt suggests in her seminal work on the state of modern humanity, *The Human Condition*, "but the life of the species" (8). "Work and its product, the human artifact, bestow a permanence and durability upon the futility of mortal life and the fleeting character of human time" (8). Farren recognizes the inevitability and necessity of "invention" but wishes to regulate the frequency of its growth and appearance. Moore, unlike

Farren, considers the workers to be ungrateful for the skills factory work provides. For Moore, these skills may be used as human capital, to improve workers' value within a mechanized marketplace. Moore deliberately pits skilled and unskilled workers against each other as a means for exploiting human capital, and this point is further substantiated when Moore recruits loyal workers to fend off the Luddite attacks.

The gap between the ethical languages employed by Moore and Farren is substantial and reflects how industrialization prompted the loss of worker self-management, which caused the separation of labor and management within the emergent factory system. Moore and his workers fail to achieve a common ethical language but leave their individual ethical languages significantly in place as modes of differentiation. By retaining their individual ethical languages, Moore and his workers preserve their subjective interiorities but enable the possibility of a tale yet to be told. This tale is the unspoken metanarrative that will unite their stories within the context of ethical events such as the Luddite riot at Hollow's Mill. Within the ethical event lies the possibility of making choices: to differentiate one's self or to be assimilated. These are the choices used by Brontë to frame the development of crowd psychology within the novel, and they are choices to be reconciled by the novel's characters.

Crowd psychology offers a number of theories for measuring how behavior might be altered within groups. According to the contagion theory, the desire to fulfill crowd expectations becomes contagious as individuals are coerced into making choices that necessitate the performance of new roles as leaders, followers, or observers. This theory suggests that as individuals are immersed within crowds, subjective interiority is displaced by crowd objectives. However, convergence theory departs from the idea that crowds exist and are independent of their human subcomponents. As convergence theory suggests, individual identity is not merely subsumed within the crowd, it is recast according to the fulfillment of mutual expectations that prefigure behavior. In other words, individuals are not chosen by crowds. Rather, they locate themselves within types of crowd that employ a similar ethical language. Individuals form crowds that might enable their *own* speech and action and assure their *own* interests.

As *Shirley*'s narrator elaborates, "All men, taken singly, are more or less selfish; and taken in bodies they are intensely so" (2.10.166). When we think of crowds, we are not content to synonymically link typologies; rather, we seek to define and to distinguish both the slight and significant advantages we might achieve from within these groups, because to do so would be to acknowledge the significance of our own contributions to the crowd. Crowd types may therefore be employed as categories for mobilizing and arranging individuals, and these categories structure social reality.[17]

To merely distinguish normative and informal crowds according to group typologies like predisposition toward violence diminishes their connecting similarities. Within both types of crowds, there exists the potential for achieving states of equilibrium and nonequilibrium. Normative crowds sustain their equilibrium by remaining passive and homogeneous. By avoiding moral doubt, normative crowds dismiss the possibility of alteration. Their behavior is contingent on the fulfillment of expectations. Each member anticipates other members will act in ways that affirm their collective experience. Actual behavior of individuals is anticipated by the normative crowd. When individuals' actual behavior suits the requirements of anticipated behavior, crowds achieve in-group homogeneity that builds upon public categories of class and gender.[18] Individuals perform the requirements of specific roles attributed to them and are rewarded with social status which is reaffirmed within the group.

Based on the prefiguration of assumptions and fulfillment of mutual expectations, normative crowd members harbor no further tendency toward differentiation. Yet, it is always within the individual's power to reject common knowledge and to disrupt prefigurations of social continuity through individuation. Informal crowds or mobs are not able to achieve equilibrium because their human subcomponents are always in the process of individuation and adjustment. Mobs exist in a state of nonequilibrium through which individuals commit violence of the mind. Traditions and commonplace value biases are displaced within new conceptual fields of meaning and differentiation, fields that operate beyond circumstances contingent behavior might explain. Characterizations of crowds might typify collective behavior, but they can neither fully obscure nor disconnect subjective interiority from the radical behavior it makes possible.

From family gatherings to schoolyard frolicking, from playing sports to viewing films, at some point or another, every individual has found himself or herself occupying spaces within crowds. From within these spaces, individuals may learn to articulate a subjective voice, to navigate social relationships, and to ally themselves with those who might appear, think, speak, and act much like them. It is this sense of self-awareness, this subjective interiority that insulates them from the objective realities of the external world. However, recalling their activities within groups might be enough to jar their subjective interiority and cause the loss of language, the failure of speech, and the absence of identity by revealing their involvement in ethical events.

Mob behavior has been commonly described as mindless or savage. However, such descriptions diminish individuals' potentials for achieving interior movement by distancing themselves from the roles they assumed prior to mob participation. For example, individuals might speak and act in ways that fulfill the expectations of mob members, yet violate the instantiation of rules, norms, and values. Distinguishing others as mob members or nonmembers rather than as individuals causes a loss of self-awareness, or deindividuation. Through deindividuation, individuals lack accountability for speech and action. The knowledge gained from deindividuation releases individuals from prefigurations of knowledge and identity, which serve as social controls for regimenting behavior. In effect, by succumbing to mob will, individuals enable the possibility of differentiation of reality and identity.

In his seminal work, *The Crowd: A Study of the Popular Mind*, Gustave LeBon considers the ways in which individual behavior might be affected by crowd psychology. LeBon theorizes how stages of anonymity, suggestibility, and contagion might direct individual behavior toward a collective experience. As LeBon concludes, "When, however, a certain number of individuals are gathered together in a crowd for purposes of action, observation proves that, from the mere fact of their being assembled, there result certain new psychological characteristics" (2). LeBon's idea of collectivism versus individualism suggests that

> [i]n the crowd the collective mind takes possession of the individual. As a consequence, a crowd member is reduced to an inferior form of evolution: irrational, fickle, and suggestible. The individual possibly controlled by the crowd's leader, and capable of performing any act, however atrocious or heroic. (Postmes)

According to LeBon, the individual is deindividuated, that is, experiences displacement of self through group participation and becomes predisposed toward antinormative behavior. However, LeBon does not suggest how these stages of crowd participation might be misappropriated by individuals to achieve interior movement between psychological states of individuation and deindividuation, movement which might assure radical differentiation. If differences of reality and identity cannot be articulated and competing narratives cannot be harmonized, how, then, may ethical events be represented and conceptualized? By bridging the gaps between micronarratives and metanarratives, we may analyze and evaluate how crowd and mob participation might be used as a strategy for redefining collectivism and individualism in relation to the ethical event.

In *Shirley*, Brontë destabilizes subjective interiority to emphasize the loss of identity within crowds as a terminal stage of deindividuation. Brontë's characters of Shirley and Robert consider and experience how their identities may be retained or accentuated within crowds and achieve full plenitude within its plurality. Robert's integration of new technologies of production within the mill increases the efficiency of labor but also decreases the need for workers. The introduction of new machines within the mill unsettles the workers, who protest conditions of unemployment and privation inspired by new technologies. The workers' demonstrations eventually turn volatile, and they employ violence as a means of expression or speech for assuring alteration. To consider the workers' violence as not just a physical act but as a mode of linguistic expression is to avoid the convenient separation of mind and body from those whose physical disembodiment from labor became the metaphorical representation of industrial progress.

Individuation necessitates that mob members continually readjust themselves to new expectations. When an individual alters his or her behavior, he or she affects other individuals, who alter their own behaviors in kind. It is through this process of readjustment that differentiation is assured and left as a supplement to the organic unity of crowds as a whole. Demonstrating "who" we are is as important as recognizing "what" we are and what we may yet become. We locate our identities in relation to our perceived membership in social groups. It is from within these same physical and ideological formations that we may achieve movement between cultural conceptions of language, meaning, and value by continually shifting from individual to group action and from separation to reintegration. The ways in which individuals occupy these transitional or liminal spaces suggests the success of identity transformation. In effect, the individual is never completely lost within the social group, but rather is rediscovered through group involvement.

Characterizations of crowds as social and cultural phenomena have been used to craft human interaction through social exchanges of meanings and values. As individuals, we derive our notions of identity from three levels of abstraction: personal, social, and human. Each level of abstraction bears its own psychological and emotional resonance and amplifies individual sensations of human experience. However, it is through the social level of identification that humans acquire plurality within liminal spaces. It is within liminal space that identities are reorganized and reconstituted and humans are understood to be always in the process of becoming. Liminal spaces enable the possibility of radical action, but this possibility is self-reflexive. An individual may be integrated within the crowd or be radical and differentiate one's self. Yet, radicalism may also be achieved from within the crowd when individuals attempt to signify themselves as altered by their crowd participation. Thus, liminal space allows the continual possibility of differentiation through radicalism. For example, while at the picnic, Caroline and Shirley observe "nearly a thousand adult spectators," but Shirley's "eagle acuteness" distinguishes Moore's "tall stature and straight port" from the rest of the crowd: "He looks amidst the set that surround him like Eliab amongst humbler shepherds" (3.6.313). Later on, Caroline mistakes Moore for the "shabby little figure of a soldier" guarding the entrance to the mill. In another instance, Shirley easily distinguishes Moore from the loyal workers and friends who gather about him and await his commands (3.8.342). Shirley's ability to recognize Moore, both at a distance and from within crowds, might suggest how radicals signify identity through behavior.

Shirley, much like Moore, stands out from the crowd and stands against the crowd, specifically the unruly workers marching on the mill. Like Moore, Shirley charts a path of individuation that distances her from familiar gender roles (3.1.209). Her confidence and courage parallel Moore's own attributes. However, it is her possession of wealth, which she uses to save Moore from financial ruin, as well as her counsel that places them on a more even footing as radical subjective individuals. Yet, such radicalism is not without its limits.

When Shirley and Caroline leave for the mill, Shirley is well aware of the impending conflict and arms herself with a pistol. Shirley assumes the role of protector, a position expressive of male domination, possession, and control of female subjectivity. Although Shirley "would have given a farm of her best land for a chance of rendering good service," she is made impotent within a moment of clarity. As Shirley explains to Caroline,

> Instead of amazing the curate, the clothier, and the corn-dealer with a romantic rush on the stage, we stand alone with the friendly night, its mute stars, and these whispering trees, whose report our friends will not come to gather. (3.8.343)

Rather than engaging in conflict, Shirley removes herself (and Caroline) from the action and observes the attack on the mill as spectacle. The presence of the rioters reminds us that even in reticence, Shirley and Caroline become observers, passive participants to the riotous events that will ensue. As witnesses to the event, their crowd involvement is confirmed, as is their capacity for actualizing their subjective identities.

The potential for engaging individuals in discourse, for organizing human resources, and for enabling physical and ideological movement within liminal spaces demonstrates the power crowds exert over human interaction. However, it is the extreme crowd or, for the purposes of argumentation, the mob, not the moderate crowd, which enables separation and reintegration of self and other and ultimately establishes their unbreakable connections. It is a knowledge that assures self-preservation beyond any sense of group imperative. Within extreme crowds, individuals recognize the potential for alteration of identity and reality by differentiating themselves from others. How individuals occupy liminal spaces within crowds determines not only group dispositions but also the potential for achieving individuation of self.

Individuation should be considered as more than a transformative stage of physical or ideological differentiation. It is self-awareness and self-discovery which excite moments of social separation. However, these moments eventually yield to reintegration. In *The Social Construction of Reality: A Treatise in the Sociology of Knowledge*, Peter Berger and Thomas Luckmann argue that within the spaces of transformation,

> [t]he individual internalizes the new reality, but instead of its being *his* reality, it is a reality to be used by him for specific purposes. Insofar as this involves the performance of certain roles, he retains subjective detachment *vis-à-vis*

> them—he "puts them on" deliberately and purposefully. (*The Social Construction of Reality* 172)

Within the crowd, individuation is often attenuated by group consensus, which should be mistaken neither for submissiveness nor for voluntary action. Group members share a common perspective legitimated by normative values that prefigure their responses to collective experience. Individuals locate themselves within crowds by recognizing signs and meanings which correlate their relation to and placement within specific public formations. For example, members of a social club might recognize one another through common appearance, demeanor, or symbolic gesture. In the absence of speech, the representations assume linguistic connotations that reveal a hidden dialogue used for organizing group members. When the defenders of the mill await the impending attack, there is an implicit exchange that takes place with the rioters, a nonverbal exchange that admits a transference of power through the smoke-filled haze and volley of gunfire: "Shots were discharged by the rioters. Had the defenders waited for this signal?" (3.8.344).

Within the mob, consensus is neither voluntary nor possible. Mobs allow for continuous contestation and competition wherein singular voices may achieve moments of domination and control of human agency toward an emergent event. Mobs, unlike crowds, lack the potential for implementing consensus as a strategy for organizing and directing collective action. Imagine, for a moment, a group of children at play when one child initiates an antagonistic and personally pleasing activity: tossing rocks. The organizing principle of the children might be to habitually entertain themselves through carefree diversions, games, projects, experiments, and so on. But, by redirecting daily routines and activities, the organizing principle is placed under erasure to better suit the interests of a single child. It is the individual leader or, more precisely, the agitator who vocalizes the need for group participation and cooperation, only to usurp group solidarity by instilling in each group member the volatile impulse toward individuation. The agitator may be forgotten, but the impulse is not. Thus, as each individuated self is subsumed within the crowd, the group impulse grows stronger, and each passing inclination or injunction is inhibited by group control.

When Moore fortifies his mill, he exhibits a composure that belies his preparedness. As "he stood to the defence with unflinching firmness; those who were with him copied his spirit, and his demeanor" and embodied Moore's impulses (3.8.345). Much like the children following the impulse of play, Moore's defenders engage in role-playing by mimicking Moore's behavior. The preciseness of their performances is a mode of resistance to the nonnormative behavior of the rioters, who "had never been so met before … [and] was what they never dreamed of encountering" (3.8.345).

Each time a group member offers an alternative suggestion for a group activity, the group impulse countermands the suggestion and redirects group movement toward the event horizon. To return to an earlier metaphor, as the children become more disagreeable, the power of suggestion, which helped to collectivize them, is now diminished. To accept consensus is to avoid doubt and to fear the dangers of difference. But the children's fears of group consensus instigate radical subjective individualism. Rather than attempt to accommodate one another, the children dismiss one another's interests and give preference to their own orientations and rationalities. In the absence of routines, procedures, objectives, and consensual logic, the children fragment into individual pockets of activity which occupy common liminal space.

The pockets of activity are not the event, but rather prefigure its conceptualization. As the group members' enthusiasm ebbs and flows and reshapes the dynamics of each pocket, there is not any clear symmetry or division of activity between them. The security of the familiar is supplemented by an acceptance of activity as event, and it is within this transitional stage that liminal spaces are used to reconstitute subjective identity.

The children are only perceivable as a social formation because they share physical space, yet they urge toward ideological differentiation. Collective movement is achieved as each child correlates his or her individuation as an advantage of participation within the emergent mob and acts to assure its continuance. It is from within this social formation of disharmony, disunity, contestation, and competition that mob psychology takes hold and replicates prevailing modes of capitalist production.

When an individual blends into the crowd, the individual enjoys freedom of anonymity through which he or she may reinvest speech and action with a sense of enthusiasm and urgency that "ferments in the destructive force of feeling[s] that [must be] stated."[19] But, anonymity is acquired at a high price. The vocalization of a crowd member's thoughts does not provide the opportunity for contradiction and contradistinction. Rather than voicing differences, crowd members yield to the demands of group harmony because they absorb the fallacy that individuals may achieve empowerment through group participation. Their individual interests become strengthened and enlarged by the spectacle of the crowd. However, as the individual is assimilated by the crowd, subjective interiority is displaced but not lost.

The displacement of subject is supplemented by group imperatives that fabricate a false sense of solidarity. Distancing oneself from collective thought and considering one's differences from the crowd do not free the individual from crowd psychology.

Nor does rebuking commonality and consensus allow individuation from the collective will. The process of individuation requires speech to restore subjective interiority and excite the behavior of the crowd from reactive to proactive, from compliant to disobedient, from consensual to disagreeable. As each crowd member exceeds individuation, vocalization becomes a strategy for accentuating individual identity and for securing private interests. As the rioters assault Moore's mill, "[t]he hitherto inert and passive mill woke: fire flashed from its empty window-frames" (3.8.344). The volley of gunfire between Moore and the rioters mimics linguistic exchange, with violence assuming the vocative mode. Within the exchange, power is conferred:

> They heard the rebel leader cry, "To the back, lads!"
> They heard a voice retort, "Come round, we will met you!"
> "To the counting-house!" was the order again.
> "Welcome!—We shall have you there!" was the response. (3.8.344–345)

"In [Moore's] voice, the yell of the rioters is matched and undone" (Plotz 175). Speech simultaneously becomes the signifier of individuation and what is being signified by collectivization as its signifier. The crowd is, therefore, the symbol. Violence and speech are acts that give meaning to the symbol.

When the mob of Luddite rioters, which had been previously glimpsed, makes its presence known in the blackness of night that engulfs Moore's mill, Shirley and Caroline witness the assault from a distance as "a crash—smash—shiver stopped their whispers" (3.8.343). Crowds occupy static space as witnesses to ethical events. But, by witnessing the event, might Caroline and Shirley be held any less accountable for their involvement in its formation? Their import to events is limited by prefigurations of knowledge that make consensus not only achievable but an essential part of social order. Their perceptions of the ethical event correspond to normative values, attitudes, and behaviors which precede the event. Caroline and Shirley observe the event but do not participate in actuality. Rather, they conceptualize the event in terms that substantiate present social order.

> Both the girls felt their faces glow and their pulses throb: both knew they would do no good by rushing down into the mêlée: they desired neither to deal nor to receive blows; but they could not have run away—Caroline no more than Shirley; they could not have fainted; they could not have taken their eyes from the dim, terrible scene—from the mass of cloud, of smoke—the musket-lightning—for the world. (3.8.345)

Although crowds may be transformed by an event, Caroline's and Shirley's experiences of the event are preconditioned and influence conceptions of its outcome: "'How and when would it end?' was the demand throbbing in their throbbing pulses. 'Would a juncture arise in which they could be useful?' was what they waited to see …" (3.8.345). Rather then choosing a side, they merely recognize prefigured signs and meanings revealed within the context of the event and await its inevitable conclusion ("they quickly stepped aside into a shed … [from] whence they could see without being seen" [3.8.346]).

Contrarily, mobs occupy the dynamic space of a moment that awaits signification. Their emergent behavior eschews the stasis of normative crowds and enables them to signify their relation to the event. Once individuals commit themselves to a particular behavior, the behavior becomes a catalyst for galvanizing group participation. This is demonstrated by the yell that echoes through the moors preceding the attack at the mill and becomes a clarion call for riotous actions throughout neighboring industrial communities.

> A yell followed this demonstration—a rioters' yell—a North-of-England—a Yorkshire—a West-Riding—a West-Riding-clothing-district-of-Yorkshire rioters' yell. You never heard that sound, perhaps, reader? So much the better for your ears—perhaps for your heart; since, if it rends the air in hate to yourself, or to the men or principles you approve, the interests to which you wish well. Wrath wakens to the cry of Hate: the Lion shakes his mane, and rises to the howl of the Hyena: Caste stands up ireful against Caste; and the indignant, wronged spirit of the Middle Rank bears down in zeal and scorn on the famished and furious mass of the Operative class. It is difficult to be tolerant—difficult to be just—in such moments. (3.8.343–344)

Although mobs lack the collective intellect of crowds, they allow for individuation and plurality. Mobs lack group psychology, yet they enable the actualization of individualism. They are comprised of individual personalities exhibiting polymorphous characteristics which enable the possibility of public action by legitimizing private concerns.

Mobs, like normative crowds, engage in social exchanges prior to the event. Their mind-sets and actions are predetermined, and they merely await the event as an opportunity for alteration. Their plurality of voices enables individuation while simultaneously becoming a cacophony that obscures motivations and purposes. By reacting to the voices of all members, the mob accedes to none.

The same freedom of anonymity that reduces identity also necessitates the absence of individual accountability and enables participants to share innocence and deny guilt. When one acts as a member of a crowd, one's interests are shaped and directed by the collective will. Thus, the crowd reduces individuality and abstracts self-identity to assure its internal functions. What emerges from this process is a collective consciousness of intersubjectivity that is irreducible to individual interests or moral responsibility.

The significance of the ethical event might be further elaborated by comparing it to Alain Badiou's theory of the event in *L'Etre et l'Evénement and L'Ethique.* Badiou's concepts stress radical departure from the status quo (i.e., prefigurations) of a particular situation. Mobs exhibit radical behavior and initiate actions that challenge law and authority and disrupt social order. By establishing nonaggression as the defining characteristic of the normative crowd and aggression as the defining characteristic of mobs, readers may respond to them as passive agents of compliance or active agents of individualism, respectively. However, these notions fixate on the depiction of mobs as violent entities. Mobs are volatile and do maintain high levels of enthusiasm, and although violence is achievable, it is not the mob's intended purpose. Rather, it is merely a strategy of engagement that must eventually yield to reconciliation.

Radicalism found in industrial novels, therefore, does exact a heavy price, and that price is the displacement of individual identity within crowds that seek to contain and to assimilate radical characters. Industrial novels record "features of the era's crowds that no other historical sources can supply," and suggest how the crowds became the novelists' "potent rival to the representational claims [of literature]" (Plotz 2).[20] In her novel *Shirley*, Brontë offers readers social antagonists like Shirley and Robert, antagonists whose thinking eschews conventional wisdom in favor of radical reform.

In *Dialogues*, Gilles Deleuze concludes that Thomas Hardy's characters "are not people or subjects, [but rather,] they are collections of intensive sensations" (Deleuze and Parnet 39–40). Deleuze observes in Hardy the "individuation of subject" or depsychologized character (40) who is made resplendent by experiencing the external world through various sensory modalities. Much like Hardy's characters, Brontë's character of Shirley becomes the agent of sensory interaction with the external world. Unlike Eliot's characters, who strive to transcend their social and cultural limitations, Brontë's characters are material objects acted upon by the external world, a world experiencing cultural and economic transformations.

During the Victorian age, industrialization necessitated the intermingling of classes and genders which, until that time, endured differentiation and segregation. Crowds demonstrated the "infinite connectability" of the modern industrial community (Munro 132). London's economic growth, for example, within the early part of the nineteenth century expanded the opportunities for social interaction, which "meant that chance encounters on [its] streets produced a new sort of social life, both a pleasant and a threatening urban anonymity" (Plotz 1). But, anonymity came at a high price. Crowds became the "visible manifestation of an increasingly incomprehensible city … always supplemental, always understood as a form of excessiveness and/or superfluousness" (Munro 1). Crowds were "staged as audiences" to the growth of technologies of production and the resulting social and economic changes these innovations inspired (Munro 11). The presence of crowds as spectators to technical processes helped to legitimize the economic rationality of technologies and to reinforce the representation of how crowds were formed in urban settings.

Industrialization broadened the avenues of social intercourse and initiated contact between classes and genders within newly forming public spaces, spaces that infringed upon individual privacy and altered Victorian conceptions of identity. Periods of severe overcrowding and forced human displacement within British industrial cities exacerbated ideological confusion and created ambivalent spaces where traditional rules of discourse and identity were no longer viable. It is within "those ambivalent spaces, [that] the symbolic oppositions to middle-class ideology—inside and outside, private and public, domestic, and economic, do not hold; as a result the lines that demarcate various components of identity—social, sexual, racial, familial, professional—begin to waver" and make social alteration inevitable (McCuskey 424).

The interactions of distinct social groupings inspired the emulation of class and gender traits. The manipulation of representations of identity within Victorian England took place from within crowds and altered their social composition. Crowds did not deindividuate; rather, they enabled differentiation of identity and social mobility. For example, women laborers demonstrated the "hubris of economic man" (Watt 95) by replicating masculine appearance and demeanor, because Victorian society made no place for the economic *woman*. Thus, women laborers exhibited masculine characteristics (e.g., appearance and demeanor) in order to secure the economic individualism that was made possible to men and not to women. Crowds enabled individuals the opportunity to engage in role-playing by obscuring their "natural" identities. Role-playing disrupts the process of signification and prevents reactivation and affirmation of the lost subject:

> What the lost other and the historical past have in common is the quality of radical otherness, an alterity that resists assimilation by either the mourning subject or social present. To recognize alterity, at least in a certain sense, can appear as a dangerous and disruptive force lodged within the self and the subject's experience of the social.[21] (Clewell 131)

For Shirley and Caroline, the lost subject is located in their moments of self-awareness, their moments of full subjectivity, their moments of complete linguistic failure, when their observations of the Luddite riots do not support a common ethical language. Shirley and Caroline distinguish their conceptions of the ethical event through different ethical languages. Caroline internalizes the experience of the ethical event and affirms her roles and obligations to the social order. Shirley interprets the event through her ethical language, a language that differentiates her from Caroline and from those who seek to maintain the status quo. Berger and Luckmann argue that through the process of socialization,

> [t]he individual internalizes the new reality, but instead of it being *his* reality, it is a reality to be used by him for specific purposes. Insofar as this involves the performance of certain roles, he retains subjective detachment *vis-à-vis* them—he puts them on deliberately and purposefully. (*The Social Construction of Reality* 172)

The repetition of performative acts of role-playing fulfills the requirements of social identification, if only in a limited sense. Crowd members successfully play roles, but mimicry of each identification further distances them from their familiar orientation.

The possibility for nonnormative acts within society may undermine the stability of identity categories by making performance imprecise, but it may also enable a greater degree of precision in the performance of roles as a form of resistance to nonnormative acts. Thus, the use of role-playing might serve as a paradigm for evidencing how Victorian conceptions of self were being advanced in early modern Europe. Role-playing allows crowd members to identify and anticipate possible problems and reveal fears and anxieties people have about an ethical event or moral action. Role-play fosters group and individual confidence and competence by allowing for the possibility of power exchanges.

Shirley experiences how different attitudes and behaviors inscribed in social classes establish a tension between utopian ideals and the conventions they put into question. She struggles to redefine her identities in response to cultural transformations accelerated by British industrialization during the Victorian age. Shirley attempts to rationalize the consequences of her crowd involvement in order to reconstitute identities that would inform and sustain the self.

It is, ironically, the crowd that cultivates the individual's desire to reclaim and to reaffirm the self. From within the crowd's cacophony of voices, the individual may achieve a calming singularity of silence. It is through the unspoken moments of silence that the individual is able to restore a sense of subjectivity, a sense of distance from the crowd as both a physical formation and a psychological abomination. Through a process of differentiation, Shirley struggles to acknowledge the connections between self and other and between individual and crowd. To transcend physical and ideological boundaries and reclaim her subjective interiority in relation to the present culture, Shirley must both participate in the crowd, then distance herself from her crowd involvement, not through outward public mobility but through private inward movement.

For example, Shirley and Caroline Helstone, cousin and bride-to-be of industrialist Moore, lead their Sunday-school class along a narrow lane (Royd Lane) on their way to a picnic spot and encounter a rival class obstructing their movement to the far end of the lane. Shirley asks Caroline, "Is it our double? … our manifold wraith?" (2.6.303). Caroline muses, "If you wanted a battle, you are likely to get one,—at least of looks" (2.6.303).

The tension between the groups of children previews the forthcoming battle at Hollow's Mill (owned by Moore, whose mechanization of production instigates discontent among the working class). It also reestablishes the idea that crowd participation plots coordinate points of reference for assuring intersubjectivity by linking people through distinct social, cultural, and economic locations.

Shirley's comments about the encounter are filled with latent meaning and suggest how the struggle either to express or to deny aggression forces either one's "retreat from the world, direct confrontation with it, or a partial break with the conventional bonds evincing membership in it" (Lane 201). Although Caroline and Shirley share a common disinterest of "[others] different from me" (1.7.114), it is Caroline who comfortably retreats behind her group of students "to find protection in their company [rather] than to patronize them with her presence" (3.6.308). However, Shirley's comments suggest her desire to confront the other in order to achieve self-affirmation by distinguishing herself and her group from the "Dissenting Methodist schools, the Baptists, Independents, and Wesleyans, joined in unholy alliance, and turning purposively into this lane with the intention of obstructing our march and driving us back" (3.6.303). The groups of students, jockeying for position on a narrow lane, become figurative representations of the conflict between self and other and also represent how both differentiations and connections shape the relationship between individuals and crowds.

Hidden from the crowd, Shirley and Caroline cast a far-off glance at the workers' attack of Hollow's Mill and struggle to differentiate between conceptions of crowd and mob as not only theoretical abstractions but also physical formations which represent how "anxieties about the potential lawlessness of crowds implicate[d] Victorian anxieties about the lawlessness of the novel form itself."[22] It is Shirley's and Caroline's "consciousness [which] channel[s] the novel's account" (Plotz 167). Shirley and Caroline witness the attack on the mill and begin to correlate the actions of the workers and the mill owner as inseparable, despite the "class, vocational, and even dispositional differences" between them (Plotz 173). Without Moore's mechanization of labor, there would not be a workers' revolt, but without the revolt, there would not be any need for reform. Even after Moore reaches an agreement with the workers, the hostile crowd remains a looming possibility that may be realized in the not-too-distant future of postmodern technological progress.

> The connection between Chartist claims to represent the working class and *Shirley*'s claim to represent the Yorkshire Luddites helps to clarify ways in which the era's literature is produced and [circulated] within an avowedly contestatory public sphere, within which various discourses vie to mandate what might count as public action and what as circulatable private feeling. (Plotz 155)

Brontë crafts the character of the crowd as a theoretical and prototypical persona to embody values and needs that would resonate with individual citizens and collectivize their personal interests within an emergent working-class agenda. However, such an agenda is continually reshaped in service of technical progress. The rhetorical power of cultural, political, and social dialogues is filtered through an ethical language that supports technical innovation and economic expansion at the expense of social continuity. Crowds work within the realms of discourse to appropriate knowledge and foster intersubjectivity. However, this matrix of intersubjectivity has been misappropriated and transformed into a technical apparatus for manufacturing consensus and for regulating language. The control of language becomes even more poignant at the conclusion of *Shirley*, when the Luddite mob is reassimilated within the same technical structures it sought to destroy. Inevitably, the social reforms emphasized within Brontë's novel might be better termed as a technical reform, which acknowledges how linguistic structures of enclosure have been used to prescribe the limits of discourse within an evolving technical sphere of rationality.

Chapter 4

Tracing the Absence of Place-Identity in Eliot's *Felix Holt*

> The philosophy one has does not depend solely on the kind of person one is. It depends more essentially on the time in which one lives and, above all, the way in which one *belongs* to the time. (emphasis added)
>
> —Ernst Bloch

Ernst Bloch's comments on modernity suggest a materialist position for existing in the Now. They also allude to the ways in which traditional forms of history make identity unreliable by interweaving chronology with personal narrative to produce an ontological singularity. The concept of simultaneity was used by Bloch to refer to the ways in which people may live in the same location at the same time, but not in the same time.[23] Different attitudes and behaviors inscribed in social classes establish a tension between utopian ideals and the conventions they put into question. George Eliot explores these concepts in *Felix Holt* by employing modern themes like the conflict of self in society within her fictional work.

Felix Holt, like *Shirley*, is set against a tumultuous backdrop: Treby Magna's election day riots of 1832. The character of Felix Holt struggles to link fragments of his past to define a new sociospatial identity in response to cultural transformations accelerated by British industrialization during the Victorian age. Eliot appropriates the opposition of self and other and spatially reconstitutes it in terms of the binary opposition between conceptions of "historical" and "personal," "public" and "private," and "outsideness" and "insideness" as representations of belonging to the present culture within her novel. These oppositions are further qualified according to perceived physical attributes such as gender, sexuality, morality, and class. Felix's conceptions of morality and class as performative acts suggest how sociospatial identity is inscribed according to referents that determine his proximity to historically prescribed definitions of "insideness" used to designate placement within the social hierarchy. I will focus on how class expectations and habitus inform the construction of spatial identity within public and private spaces that are being redefined according to the demands of industrialization. In suggesting this cultural and psychological construction of identity, I wish to stress the interplay of "complex strategies of cultural identification and discursive address that function in the name of 'the people' or 'the nation' and make them the immanent subjects of a range of social and literary narratives" or inscriptions that limit the potential for individual agency (Bhabha 30). By denying class distinction, Felix hopes to transcend socioeconomic classification and define himself in his own terms, terms that locate meaning in past culture. In order to transcend himself, Felix must first transcend bourgeois culture and subjective interiority by producing anarchy, dissolution, and disbelief within himself. Yet, even while defining himself in opposition to class expectations and habitus, Felix still works within the parameters of signs and significations that inscribe his identity.

Class stratification serves as the inspiration for fostering a sense of community that would unify disparate social groupings within an emergent national consciousness. Teasing out the ways in which signs produce fissures between these categories may demonstrate how individual agency might have been reempowered in Victorian society. By negotiating his conceptions, interpretations, and beliefs about his relation to his cultural and psychological landscapes, Felix produces a continuous slippage between these categories. The slippages cause him to trace the absence of a place-identity that would inform and sustain the self and allow him to fulfill his desire to create a personal narrative written in the context of "belonging" to the present culture and sharing in "the labour and common burthen of the world" (27.258).[24]

Concepts like "the determined universe, the nature of moral responsibility, and the importance of human community" affect Felix's experiences of self-identity and his relation to others (Galloway). His human experiences are essential for producing his self-awareness and are the basis for all other experiences, so by struggling to locate the "we" experience, the "I" experience eludes Felix. He experiences movement of supplementarity between conceptions of presence and absence of belonging to the present culture, which informs his history of being. Derrida's notion of the trace left by the absent sign in the process of signification might be useful in justifying Felix's desire to locate a suitable place-identity that would accommodate both public obligations to the working class he chooses to represent and private concerns for his relationship with Esther Lyon, daughter of Reverend Rufus Lyon.

Although in *Of Grammatology*, Derrida adapts *trace* as a theoretical term in relation to a number of "nonsynonymic substitutions" (e.g., différance, arche-writing, and spacing), it will be beneficial to isolate trace as a by-product of the presence-absence opposition within the context of Eliot's novel. It will also be necessary to consider classism as a convenient metaphor for linking Felix to a shared history that treats "a people, a nation, or a national culture as an empirical sociological category or a holistic cultural entity" (Bhabha 140). Felix's inability to maintain a dialogue with his past denies him the knowledge necessary for creating a personal narrative. However, as Derrida suggests, the absence of a thing leaves a residue that indicates presence. Thus, even as Felix struggles to define himself, he is always trying to restore what he has lost: a place-identity that assuages his sense of belonging to the present culture.

There are numerous works that allude to the themes of memory, presence, and absence and are available for inclusion within my essay.[25] However, I will concentrate my investigation on selections from critical works that incorporate the themes as supporting corollaries to my analysis of Eliot's novel. Concepts of self and other, spatially reconstituted in terms of presence and absence, suggest Felix experiences "belonging" or "insideness" not as an absence, but rather as a presence which has not actually been present in his life. Presence, therefore, presents itself in terms of multiple histories of being that necessitate a unifying narrative, much in the same way distinct social and political groupings inform national consciousness. Without the unification of narratives, full plenitude is not achievable.

The misapprehension of significations of identity causes a rupture in Felix's personal narrative. He is displaced because he is not willing to comply fully with the prejudicial criteria of classism used for redefining sociospatial identity within the emergent culture of industrialized England. He seeks traces of his identity within the radical counterculture he chooses to represent, a counterculture which will evolve into the bourgeois majority he sought to disempower. When Felix says, "I'm a Radical myself, and mean to work all my life long against privilege, monopoly, and oppression" (16.182), his declaration works against the interests of the privileged class to which his education should have allowed him entry. His self-naming enables him to locate himself within the social hierarchy rather than allowing himself to be situated by others. It further demonstrates his willingness to displace himself from his own cultural orientation in order to fulfill his sense of moral obligation to the people of Treby Magna.

Felix rejects class expectations and habitus that would better enable him to locate a place-identity. He criticizes conventional clothing that he believes might "throttle" and restrain him in "straps" (10.129). He eschews the allure of women who might corrupt the purity of his principles: "Men can't help loving them, and so they make themselves slaves to the petty desires of petty creatures" (10.129). Although "a bachelor's children are always young … with a chance of turning out good," Felix denies himself the right to create his own legacy because children might become more of a burden than a blessing (22.232–233). Felix's unwillingness to root himself in relationships and his inability to perform cultural rites of passage (e.g., fatherhood) limit his ability to represent the people of his society, who regard him as less a leader than a "young Ishmaelite" returning from his "lawless life of the desert" (37.363).

The ways in which Felix flaunts his disregard for tradition are contrasted with his mimicry of the appearance and rhetoric of his political contemporaries who galvanize support for private agendas enacted within the public forum. Felix's mimicry also suggests how nobles of the time had begun to reject the roles and duties that defined their positions in society. It further indicates how industrialization precipitated the deterioration of a social order dependent on the ability to locate identity within a hierarchy that remains constant and immutable. Civic virtue had been realized by people fulfilling their roles and obligations within the hierarchy. However, the possibility for nonnormative acts undermines the stability of identity categories by making performance imprecise. Thus, role-playing might serve as a paradigm to suggest how changing conceptions of self were being advanced in early modern Europe, representations that signify the birth of modern individualism.

The status of cultural representation is deterritorialized as Felix becomes the embodiment of a counterculture that rejects the idea of a single symbolic order under which exclusive membership might be contained. How Felix reorients himself in response to such knowledge determines either his success or failure in signifying place-identity as a "substructure of the [historical] self" (Dixon and Durrheim 32). However, Felix gradually divorces himself from the values of the dominant culture because he is interested in recording and observing their activities to confirm the superiority of the subjective self over the other consciousnesses of the objective world: "Because you can't see my history or my nature … . I have to determine for myself, and not for other men" (27.258). Although Felix's repositioning and redefining of boundaries of identity underscore the power of self-determination as a rationalized countermeasure to historical determinacy, Felix works within the same systems of signs and significations he hoped to disrupt and devalue. Felix's radical politics merely allow the substitution of one set of expectations and habitus for another, without actually advancing dramatic social change or reform.

In her essay "The Casuistry of George Eliot," Shirley Galloway revisits Baruch Spinoza's conceptions of "adequate ideas" and "feelings" and correlates them with the ways in which Felix's attitude and behavior convey a sense of ideological symmetry and freedom:

> Spinoza's reliance on "adequate ideas" as the foundation of a personal ethical system is seen clearly in Felix Holt's *Address to the Working Man*: He does not want to reform institutions, but wants to change men's ideas. As already shown, Eliot's education of her characters through experience is a Spinozian credo. Experience and deduction are the methods of developing "adequate ideas." Through "adequate ideas" and intuition humans achieve Spinoza's idea of "freedom," another way of expressing the existence of free will in a determined universe. [R.T.] Jones observes that in Eliot's novels "... as in the life around us, some people appear to be far more 'free' than others." Yet, both Spinoza's and Eliot's idea of freedom consists of a larger understanding of oneself in the context of the world at large, and the willingness to place oneself in alignment with it; it is not an arbitrary exertion of the will in contradiction or neglect of the rest of humanity and the universe. (90)

However, by individuating himself, Felix fails to realize he has activated a sequence of events that will eventually impose *his* will on the people of Treby Magna and perpetuate, rather than diminish, the power of historical determinacy. Rather than uprooting traditional political entities, Felix's radical stance further entrenches them as immanent subjects of Treby Magna's social narrative.

Felix dismisses the "ordinary Christian motives of making an appearance and getting on in the world" and chooses "to live in a way that would prevent any one from classing him according to his education and mental refinement" (22.227). Although Eliot aligns her novel more closely with "high culture narratives,"[26] the introduction suggests how cultural and psychological landscapes offer a more complete narrative perspective of "English life … [and] labours" (4). Angela Miller suggests how

> [the] … imagined community … of the nation … rests on social and cultural abstraction … [so that its] image of nature … is framed by ideological requirements far beyond the task of view-makers. It [is] informed by viewing habits linked to specific identities and by … ambitions to create and profit from a public with their works. (207)

Working classes are presented not only as physical forms but also as symbols of reduction and abstraction. London's economic growth within the early part of the nineteenth century expanded the opportunities for social interaction, which "meant that chance encounters on [its] streets produced a new sort of social life, both a pleasant and a threatening urban anonymity" (Plotz 1). Industrialization broadened the avenues of social intercourse and necessitated intermingling classes and genders within newly forming public spaces.

Eliot uses the character of Felix to mend the ideological ruptures of modernization by integrating elements of knowledge from his cultural and psychological landscapes into an operative system of *self-knowledge* that reflects the social and moral concerns of the present:

> "Because you can't see my history or my nature," said Felix bluntly.

> "I have to determine for myself, and not for other men. I don't blame them, or think I am better than they; their circumstances are different." (27.258)

Felix's self-knowledge threatens the status quo by placing him at the center of historical reform (i.e., the passage of the First Reform Bill[27]), orchestrating his observations to maintain the subjective self. Felix's unwillingness to fulfill the social expectations of either the dominant culture or the emergent counterculture prevents his assimilation, denies him a sense of belonging or "insideness," and necessitates not only his ideological and physical expulsion from the dominant culture but also the revision of his personal narrative, which he experiences as "that mixture of pushing forward and being pushed forward, which is a brief history of most human things" (33.317).

"Social insideness," to borrow from G. D. Rowles, is a mode of place identification that enables one to acknowledge "integration within the social fabric" (301). In their essay "Displacing Place-Identity: A Discursive Approach to Locating Self and Other," John Dixon and Kevin Durrheim further explicate how one's sense of "social insideness" verifies place-identity: "Belonging, it seems, designates not just a phenomenological sense of one's place in the world; it is also an ideological location from which the *in-place* and the *out-of-place* can be diagnosed" (39, emphasis added). The "fabric" of intersubjectivity to which we may lay claim uneasily weaves tensions between insideness and outsideness. Social antagonisms begin to unravel place-identification, which is located outside of the material bounds of self and yet used to define that interior. Thus, "one could unravel the [totality] of signs in any culture by beginning with an isolated sign [e.g., identity] and following its myriad of connections" until the cohesiveness between signs is ruptured.[28] Felix experiences this rupture of signs as he is unable to adjust his language to suit the demands of the mob that confronts him on election day, along with the jury that sits in judgment of him for his involuntary act of manslaughter against Constable Tucker.

Peter Berger and Thomas Luckmann contend that "society determines the presence but not the nature of ideas" (*Social Construction of Reality* 8). Felix develops a system of self-knowledge in "isolation from the … constructs of reality" (102) in order to negotiate the new sociospatial identities that confront him. Felix crafts his observations to satisfy definitions of truth and belief that support his system of self-knowledge, which is designed to affirm his ideological purity, authority, and superiority, especially in relation to members of his culture.

Communities change and societies rise and fall, but there is a fundamental system of self-knowledge that operates within the subconscious and affects the unconscious. It is a knowledge that assures self-preservation beyond any sense of cultural or moral imperatives. Felix experiences ideological displacement by expressing a conscious will to not participate in the greed of an emergent industrial society:

> I do choose to withdraw myself from the push and the scramble for money and position. Any man is at liberty to call me a fool, and say that mankind are benefited by the push and the scramble in the long-run. (27.258)

His obligation to the present, to "people who live now and will not be living when the long-run comes" enables him to detach himself from concerns of social advancement (27.259). However, the pursuit of money and power is an outgrowth of a central pathology of modern societies' desire to colonize the lifeworld by reconstructing and reinscribing it within interlocking systems of knowledge. Felix is cognizant of the process of signification and chooses to reject the adornments of wealth and status in order to restore the dignity of the common man:

> It all depends on what a man gets into his consciousness—what life thrusts into his mind, so that it becomes present to him as remorse is present to the guilty, or a mechanical problem to an inventive genius. There are two things I've got present in that way: one of them is the picture of what I should hate to be. I'm determined never to go about making my face simpering or solemn, and telling professional lies for profit; or to get tangled in affairs where I must wink at dishonesty and pocket the proceeds, and justify that knavery as part of a system that I can't alter. If I once went into that sort of struggle for success, I should want to win—I should defend the wrong that I had once identified myself with. I should become everything that I see now beforehand to be detestable. (27.259)

Even by choosing a more "radical" course of action, Felix must operate within the framework of the political and cultural systems of knowledge he seeks to challenge and reposition these historically contingent systems within himself in order to achieve the ideological symmetry and certainty of place-identity. Felix is interested in living in the now and correcting what he perceives as flaws in the present culture. What Felix fails to realize is that the new reforms he desires are not new, but are merely signs of the past reconstituted in the present.

To merely dismiss Felix's motivations as inspired by nostalgia would discredit the ways in which he yearns for progress and reforms that were anachronistic to preindustrial culture. Rather, Felix desires a permanent archive of memory that draws from personal and collective narratives to fill the gaps caused by rapid industrial change. However, in Treby Magna, the collective narrative is expressed by technologies of production that reduce the scope of history by encapsulating it within the present tense. Technologies make no rule of permanence other than their need for continuous refinement, but this process eschews temporal movements by existing in the now. Human history emits circular motion and navigates freely between the past, present, and future. It is through this movement that archives contract and expand and become responsive to human needs.

The reforms Felix seeks, political, social, or economic, are not exclusive to the Victorian era. Rather, they are issues framed within a broader historical narrative, a narrative weaved into the tapestry of an holistic human archive. It is this archive that becomes the last defense against the encroachment of technologies into the human experience. It is this archive that enables the transcendence of time and the measure of human growth by engaging in a dialogue with the past. Unlike technologies, which perfect their own obsoleteness, human archives maintain a perpetual state of relevance. Memories, experiences, and beliefs are not merely catalogued for posterity. Instead, they are correlated and contextualized to evidence the interrelation of all archival materials.

For Felix, social change suggests the need to release oneself from any historical obligation to the past by reimagining it in the present. Eliot considers how remnants of past cultural values and norms may be reinserted into society as "new" methods for achieving social progress. However, the progressive social changes Felix pursues *are* derived from the past, which is continuously appropriated and reconstituted according to new "definition[s] and solution[s] of society's problem."[29] Unfortunately, the new definitions merely reinscribe class disjunctions that cannot be adequately resolved within the changing cultural and psychological landscapes of the novel.

The failure to resolve these disjunctions has a contradictory effect on society and reveals Victorian anxieties toward our origins and systems of knowledge. It places social hierarchy in jeopardy and yet sustains it by destabilizing identities as signs in the process of signification within a given social order. In effect, destabilization leaves open the possibility for social transformation. Reconsidering the ways in which the classifications are transformed into signs in the process of signification produces a fissure between the class categories and empowers individual agency in society: in this case, the rise of social antagonists like Felix.

As Horst Ruthrof explains, "Signs are proxies not only for something else—such as things, emotions, ideas, action, the world—but also in the sense that their semantic content requires further signs in order for the transformation from syntax to semantics to take place."[30] The reliability of place-identity is put into question in the novel's introduction. As a coach passes through the countryside, a passenger witnesses the effects of scientific inventiveness, technological industrialization, and cultural modernization:

> In these midland districts the traveller passed rapidly from one phase of English life to another: after looking down on a village dingy with coal-dust, noisy with the shaking of looms, he might skirt a parish all of fields, high hedges, and deep-rutted lanes; after the coach had rattled over the pavement of a manufacturing town, the scene of riots and trades-union meetings, it would take him in another ten minutes into a rural region, where the neighbourhood of the town was only felt in the advantages of a near market for corn, cheese, and hay… . The busy scenes of the shuttle and the wheel, of the roaring furnace, of the shaft and the pulley, seemed to make but crowded nests in the midst of the large-spaced, slow-moving life of the homesteads and far-away cottages and oak-sheltered parks. (6–7)

The passenger intimates the physical contrast between old rural England, "the district of protuberant optimists, sure that old England was the best of all possible countries" (5–6). However, industrialized England reels from the impact of profound social and economic transitions:

> The breath of the manufacturing town, which made a cloudy day and a red gloom by night on the horizon, diffused itself over all the surrounding country, filling the air with eager unrest. Here was a population not convinced that old England was as good as possible … . (5)

The contrast of old world and new world, country and city, evidences how given spaces were being conceptualized and represented in the British polity.

The passenger witnesses the physical decay and transformation of the English countryside as industry consumes the natural landscape. As the coach traveler is left to ponder the reasons why "town and country had no pulse in common," one is compelled to consider the consequences such knowledge had on the adaptation and reconstitution of national identity in historical terms:

> till the agitation about the Catholics in '29, rural Englishmen had hardly known more of Catholics than of the fossil mammals; and their notion of Reform was a confused combination of rick-burners, trades-union, Nottingham riots, and in general whatever required the calling-out of the yeomanry. (5)

The traveler's impressions of the cultural landscape echo the tension between ideals of progressive industrialization and the traditions they replaced. They also suggest how distinct social groupings such as the Catholics, working class, and artisans were not adequately represented until they revolted against the conventions of the dominant technical culture. As Lenore Horowitz suggests,

> In this rural midland society George Eliot had, of course, no place for the industrialized working class or for the artisans and colliers who appear in *Felix Holt*. Perhaps this is just as well. Like most of the Victorian novelists, she did not know the working man; she saw him at a distance, and at this remove he appeared less as an individual and more as a vaguely disturbing element in society.[31]

By revealing disjunctions between the classes, Felix becomes the "disturbing element" that disrupts the present by reconnecting it to the revolutionary movements of the past. And yet, Felix enacts a quest for future social harmony. Revolution and reformation become a two-pronged strategy for rewriting social and literary narratives. The conflict between city and country, between old England and new England, establishes a backdrop for the events of the novel, events which initiate Felix's internal struggle to locate a "common pulse" linking cultural and psychological landscapes which inform his sociospatial identity. Bernard Paris applauds Eliot's efficiency in weaving unrelated decisions into the tapestry of a broader historical discourse:

> Note how many social, economic, and political conditions and personal motivations, each with its own complex history, Eliot depicts in her explanation of the election day riot at Treby Magna and its consequences for Felix Holt. Not only does she show every situation and act as the product of natural causes; she also strives to give the whole picture, the whole network of relations, past and present, inner and outer, or as much of it as possible. (31)

Eliot offers readers a simple analogy that stresses the interrelatedness of individual decisions that may collectivize as political projections:

> Fancy what a game at chess would be if all the chessmen had passions and intellects, more or less small and cunning: if you were not only uncertain about your adversary's men, but a little uncertain also about your own; if your knight could shuffle himself on to a new square by the sly; if your bishop, in disgust at your castling, could wheedle your pawns out of their places; and if your pawns, hating you because they are pawns, could make away from their appointed posts that you might get checkmate on a sudden. You might be the longest-headed of deductive reasoners, and yet you might be beaten by your own pawns. You would be especially likely to be beaten, if you depended arrogantly on your mathematical imagination, and regarded your passionate pieces with contempt. Yet this imaginary chess is easy compared with the game a man has to play against his fellow-men with other fellow-men for his instruments. He thinks himself sagacious, perhaps, because he trusts no bond except that of self-interest; but the only self-interest he can safely rely on is what seems to be such to the mind he would use or govern. Can he ever be sure of knowing this? (29.278)

For Eliot, Felix's struggle to secure a more radical identity becomes, much like chess, a convenient metaphor for expressing the interconnectedness of microscopic decisions that multiply and induce macroscopic consequences which resonate on a national scale.[32] It also suggests that men's desires, interests, and ambitions need to be governed, because their unpredictable attitudes might polarize their judgments and place social harmony in constant jeopardy. Such rhetoric is further evidenced by the traditionally ensconced politicians of Treby Magna who recognize, if not verbalize, why social change must be made in increments to allow time for smooth transitions between conceptions and ideologies.

Felix experiences the impact of his own decisions when he is confronted by an unruly mob that has been inspired by his persuasive rhetoric. By playing the roles of radical, rebel, self-conscious poet, and cultural visionary, Felix traces the subjectively perceived sense of belonging he has lost, the signification of national identity as a psychosocial qualifier:

> The mind of a man is as a country which was once open to squatters, who have bred and multiplied and become masters of the land. But then happeneth a time when new and hungry comers dispute the land; and there is trial of strength, and the stronger wins. Nevertheless the first squatters be they who have prepared the ground, and the crops to the end will be sequent (though chiefly on the nature of the soil, as of light sand, mixed loam, or heavy clay, yet) somewhat on the primal labour and sowing. (8.105)

The previous quotation suggests how the minds of common men are fertile soil for naturalizing social concepts and political ideologies. For example, common men might be easily duped into supporting the initiatives of politicians who act with intransigence toward them because they have been bred and conditioned to adhere to a political mind-set, without considering conceptual details. The quotation also suggests the need for "hungry [new] comers," like Felix, to wrest power from the same politicians in order to better represent the interests of disadvantaged classes (e.g., the poor, artisans, and tradesmen). However, Felix's tone intimates that no matter which political group wins, the lower classes will suffer neither a disadvantage nor an advantage.

Rather, they will continue to fulfill their subservient obligation to society without receiving commendations for their "primal labour and sowing," which sustains the culture that restricts their position within an immutable social hierarchy (8.105).

As the narrator correctly surmises, these commoners suffer effacement as unknown soldiers and heroes of a culture for whom they willingly sacrifice themselves:

> We see human heroism broken into units and say, this unit did little—might as well not have been. But in this way we might break up a great army into units; in this way we might break the sunlight into fragments, and think that this and the other might be cheaply parted with. Let us rather raise a monument to the soldiers whose brave hearts only kept the ranks unbroken, and met death—a monument to the faithful who were not famous, and who are precious as the continuity of the sunbeams is precious, though some of them fall unseen and on barrenness. (16.184–185)

The narrator's comments repeat how the working-class population of industrialized Britain are reduced, abstracted, and disempowered by the same society their labor expands, legitimizes, and invigorates.

> This effacement acts to continue an intellectual colonization that dislocates the voice of the other who is cut off from being a part of history—becoming that essentialized, atemporal object, displaced from the human intellectual situation. Neither knower nor interpreter, the other is merely an ahistorical inanimate observer, robbed of the ability to become more fully human because he/she is denied that gift which distinguishes human from animal—the ability to speak—and be heard and understood. (Taneja)

Felix, as the political representative of the disempowered working class, comes to symbolize the "ahistorical inanimate observer," lost in the moment of chaotic revolt that disempowers his critical voice.

The climate of transition and unrest creates the possibility for Felix to locate his purpose or function within society: to reconstitute "the distant system of things called Gover'ment" (4). It also creates a fissure between emergent class formations that either embrace or reject tradition:

> [T]he peasant still wears his old costume wherever silence and denseness, wherever traditional customs and beliefs hold sway. He stubbornly defends his economically obsolete position and is more difficult to replace by the machine than was the artisan a hundred years ago. He is more difficult to replace simply because he controls his means of production, using agricultural machinery only as an aid within the old framework of the farm and the fields belonging to it. Here we find no industrialist introducing the mechanical loom and the like, which only the capitalist can possess against economically weak artisans. This communal form of production also makes it difficult to mobilize the economic contradictions within the peasantry. (Bloch 24)

The peasantry reshape themselves into a "uniform class" that is bound to the land they inhabit. They yearn for the "return of the good [old] times," when nature was not spoiled by the process of "urbanization." Their patterns of production and consumption have been altered and bind them to a reaction that suggests nonsynchronism, literally being out of sync with the rest of the world (Bloch 24). The peasantry of Treby Magna experience disenfranchisement and alienation from not only industrialization but also from the political process. As Felix comments,

> I want the working men to have power. I'm a working man myself, and I don't want to be anything else. But there are two sorts of power. There's a power to do mischief—to undo what has been done with great expense and labour, to waste and destroy, to be cruel to the weak, to lie and quarrel, and to talk poisonous nonsense. That's the sort of power that ignorant numbers have. It never made a joint stool or planted a potato. Do you think it's likely to do much towards governing a great country, and making wise laws, and giving shelter, food, and clothes to millions of men? Ignorant power comes in the end to the same thing as wicked power; it makes misery. (2.30)

It is the peasantry's ability to reshape themselves into a collective voice that disrupts the constancy of economic categories used to restrict social mobility. Felix views himself as the one voice that speaks for the many. However, his critical inclinations lend themselves more to theory and less to a plan for gradual reformation. When Felix does speak to the class he chooses to represent, during the election day riots, his language is diffuse, and he struggles to communicate lofty concepts of reform to the poorly educated working class. Felix's idealism cannot be grounded in colloquial language, nor can it be made more conceptually accessible through common experiences. When he addresses his followers, his language transforms them into symbols of his own process of reduction and abstraction.

Felix lacks the rhetorical power of Rufus Lyon, for example, whose sermon on the need for men to become advocates of their own moral, religious, and political beliefs presages Felix's appearance as a social agitator of progress:

> Do you think there will ever be a great shout for the right—the shout of a nation as of one man, rounded and whole, like the voice of the archangel that bound together

> all the listeners of earth and heaven—if every Christian of you peeps round to see what his neighbours in good coats are doing, or else puts his hat before his face that he may shout and never be heard? (4.53)

Lyon's comments remind us of his devotion to a solitary life of piety and self-sacrifice for his parishioners. Much like Lyon, Felix attempts to maintain ideological purity by keeping his hands so clean that he must "eat turnips" to satiate his desire for relationships. However, Felix fails to realize that rhetoric cannot simply be used to disavow cultural categories (e.g., class) without reconstituting them as necessary social formations. As Georg Lukács suggests,

> [E]conomic categories must not be seen fetishistically as abstractions—as in the vulgar economics of the bourgeoisie or in Menshevism and vulgar sociology—but as immediate forms of existence of human life, forms in which the metabolism of every individual person with nature and society takes place.[33]

Felix's noble ideas for socioeconomic reform cannot be readily inserted into his present-day society and, thus, must be compromised for the sake of the common good:

> The way to get rid of folly is to get rid of vain expectations, and of thoughts that don't agree with the nature of things. The men who have had true thoughts about water, and what it will do when it is turned into steam and under all sorts of circumstances, have made themselves a great power in the world: they are turning the wheels of engines that will help to change most things. But no engines would have done, if there had been false notions about the way water would act. Now, all the schemes about voting, and districts, and annual parliaments, and the rest, are engines, and the water or steam—the force

> that is to work them—must come out of human nature—out of men's passions, feelings, desires. Whether the engines will do good work or bad depends on these feelings. (1.30)

Felix offers a functionalist idea of political institutions he finds himself working both for and against. Felix's comments insinuate that transforming myths of social change into political and economic realities is contingent on the "passions, feelings, and desires" of the common workers who must realize the need for progress, a progress that must avoid "mechanical repetition without the opportunity for reevaluations" (Horowitz 140). As Horowitz intimates further,

> [the] myth of social transition defines meaningful change as the reorientation of society towards the future rather than the past … *Felix Holt* presents an urgent plea for the present to break free from the control of the past in the definition and solution of society's problems. (132)

However, Felix is not able to fashion his idealistic dreams into a legitimate plan of reform:

> Felix's sense of social responsibility is close to a *vocation* in the religious sense, and religious imagery abounds in the novel to describe it. It is not easy for him to translate this feeling of brotherhood into effective social action. He bangs and smashes at conventional values, but he is unable to define the positive part of his social function, that *demagogue of a new sort* (27.262) he desires to [become]. (Horowitz 136, emphasis added)

The bold and ideal leader Felix envisions himself to be is suited neither to administer social change nor to regulate the psychological impulse of the unruly mob he helped to create. Rather, Felix is the ideologue that conceives of social and political transformations that require periods of conceptual evolution and maturation before they are available for intervention.

Felix experiences rebirth as his conflicting senses of self and other, insider and outsider, radical and conformist are made homogeneous within the constructs of a new identity that supplements the historically disparate parts to form a unified whole:

> Felix was perfectly conscious that he was in the midst of a tangled business. But he had chiefly before his imagination the horrors that might come if the mass of wild chaotic desires and impulses around him were not diverted from any further attack on places where they would get in the midst of intoxicating and inflammable materials. It was not a moment in which a spirit like his could calculate the effect of misunderstanding as to himself: nature never makes men who are at once energetically sympathetic and minutely calculating. He believed he had the power, and he was resolved to try, to carry the dangerous mass out of mischief . . . . (33.316)

Felix perceives of himself as a blank piece of paper without any inscription to inform his sociospatial identity, a deferral of identification supplemented during the election day riot. "The supplement occupies the middle point between total absence and total presence" (Derrida, *Of Grammatology* 202). Felix's use of supplementation enables him to place his existence in binary opposition to his nonexistence. By denying the inscription and circulation of his assumed identity among the privileged classes, Felix discovers the limitless possibilities of identities he might assume, identities that must be activated in order for him to have human experiences and engender social ties.

Belonging has been perceived as a class-inflected sign of social sophistication that has been made available for appropriation within a variety of contexts. For Felix, to fail the test of belonging does not mean to fall from social grace, but rather to transcend its prerequisites. Belonging is a mode of critical reflection that, much like Derrida's "specter," drifts "beyond the opposition between presence, nonpresence, actuality, and inactuality, life and non-life" (*Specters of Marx* 12).

"When a sense of belonging is renounced, only then can the 'other' arise, since de-stabilization nurtures the experience of supplementarity" (Derrida, *Of Grammatology* 202). It is at this historical and theoretical juncture that Felix locates ontological singularity within the transformative event of the election day riot, during which time the mob pursues and punishes Spratt, a worker who did not support their radical agenda:

> Felix went straight to this platform and stepped upon it, crying "Halt!" in a loud voice to the men behind and before him, and calling to those who held Spratt to bring him there. All came to a stand with faces towards the finger-post, and perhaps for the first time the extremities of the crowd got a definite idea that a man with a sabre in his hand was taking the command. "Now!" said Felix. (33.318)

Felix's exclamations suggest his spirited emotion and his desire to augment his power over the mob, fellows he neither knows well nor understands fully. However, on a more symbolic level, the exclamations also offer the moment of his greatest impetus, invigorating the mob while refusing to harm a common worker, Spratt, as he preserves nothing by risking everything: title, rank, and station within society. As Paris suggests, "The reconciliation of individual needs and societal forces is a major theme of Eliot's fiction. Tragedy consists, [Eliot] argued, 'in the terrible difficulty of the adjustment of our individual needs to the dire necessities of our lot'" (1.22–23). Felix's vision of social change is compromised by the unpredictable nature of the mob and the unforeseen consequences of his own actions. During a physical altercation with Tucker that results in the constable's death, Felix reduces himself as an ideological abstraction by acting out his concept of radicalism.

It is at this stage of terminal transition that Felix achieves complete freedom by having the opportunity to attempt a rewriting of his personal narrative. Affected by the mob's heightened sense of aggression and his own sense of complicity in Tucker's death, Felix attempts to reconstitute a "sense of locatedness through a symbolic process of emplotment: a form of self-creation whereby person-place relations are turned into a credible self-narrative" (Dixon and Durrheim 31).[34] When Felix appears before the court, he recalls the confrontation, and he assuages his own guilt by reducing his accountability:

> When I threw Tucker down, I did not see the possibility that he would die from a sort of attack which ordinarily occurs in fighting without any fatal effect. As to my assaulting a constable, it was a quick choice between two evils: I should else have been disabled. (46.442)

Both the signs of revolt and the signification of identity must be placed under erasure in order for Felix to experience the supplementarity of tracing self in terms of other:

> On the one hand, [supplementarity] is that of which the possibility must be thought before man, and outside of him. Man allows himself to be announced to himself after the fact of supplementarity, which is thus not an attribute—accidental or essential—of man. For on the other hand, supplementarity, which is nothing, neither a presence nor an absence, is neither a substance nor an essence of man. It is precisely the "play" of presence and absence, the opening of this "play" that no metaphysical or ontological concept can comprehend … . Man calls

> himself man only by drawing limits excluding his other from the play of supplementarity. (Derrida, *Of Grammatology* 243)

Felix experiences erasure of identity. He is not able to speak because the presence of identity has been subvocalized as absence. The concept of supplementarity may be applied to the ways in which Felix proclaims himself a "man" by challenging the hypocrisy he finds echoed in not only the words of his opposition but also in his own miscalculated machinations, particularly when he is caught up in the fervor of the vengeful mob he helped to inspire.

The mob of workers makes its way to the Debarry estate, an immense landscape which dwarves the size of the entire village of Treby Magna. As the mob enters the grounds of the estate, Felix realizes how unpredictable and unruly the mob has become, and he becomes aware that he cannot redirect its course. Felix attempts to reduce the consequences of his participation by making himself the subject of abstraction:

> The Debarrys were not of the class he was wont to be anxious about; but Felix Holt's conscience was alive to the accusation that any danger they might be in now was brought on by a deed of his. In these moments of bitter vexation and disappointment, it did occur to him that very *unpleasant consequences* might be hanging over him of a kind quite different from inward dissatisfaction; but it was useless now to think of averting such *consequences*. As he was pressed along with the multitude into Treby Park, his very movement seemed to him only an image of the day's fatalities, in which the multitudinous small wickednesses of small selfish ends, really undirected towards any larger result, had issued in widely-shared mischief that might yet be hideous. (33.319, emphasis added)

The Debarrys represent the aristocracy, whose wealth and status is placed in jeopardy by the emergent working-class mob. Felix individuates himself from the mob while also distancing himself from his own subjective interiority. Felix's comments hearken back to earlier descriptions of "ignorant power" and suggest how individuals might be caught up in a political fervor that skews their sense of morality and dismisses their radical behavior. It also reestablishes the idea that crowd participation plots coordinate points of reference for assuring intersubjectivity.

Felix does not experience a sense of commonality; rather, he experiences a sense of "in-betweenness" as he becomes both participant and observer of the election day rioting.[35] This doubling enhances his awareness of the mob's motivations while enabling him to eschew his complicity in instigating the riots. Felix realizes the mob is governed neither by passion nor belief but, rather, it is merely a group of individuals reacting to an historical event that has collectivized attitudes and behaviors and galvanized energies toward a "dramatic collision"[36] which might not provide the opportunity for social reformation. He also insinuates the ways in which traditional and radical political parties share analogous structures and similar pathologies and are as relative to each other's identities as they are to his own construction of identity.

Felix transcends the self and is able to perceive the value that he is defending from his morally elevated point of view: human fellowship and equality. However, Felix's experiences in the riot convince him that his high ideals and purity of purpose do not satisfy the demands of society, and his efforts "[will] never be known beyond a few garrets and workshops" (45.435). Thus, Felix relinquishes his role as a "detached critic of society for the role[s] of husband, father, family, provider, and teacher" and establishes the social ties necessary for locating a sense of belonging (Horowitz 141).

> As Seamus Heaney wrote,
> The places I go back to have not failed
> But will not last. Waist-deep in cow-parsley, I reenter the swim, riding or quelling
> The very currents memory is composed of, Everything accumulated ever
> As I took squarrings from the tops of bridges
> Or the banks of self at evening. (101)

Heaney's words suggest how one must correlate experiences of cultural and psychological landscapes to trace the contours of identity. Felix makes these correlations by embracing the cultural landscape, linking it to his ancestral origins, and then tracing its availability for engaging a broader, more inclusive social discourse: "I have my heritage—an order I belong to. I have the blood of a line of handicraftsmen in my veins, and I want to stand up for the lot of handicraftsman as a good lot" (27.262). Felix attempts to locate himself as the political voice of an emergent middle class which might have a more resounding influence over the direction of British history by claiming a "freeman's share" of wealth (30.289). However, Felix's idealistic dream of invigorating the national-political consciousness fails because his concepts of new reform cannot be easily integrated into the present-day, for as the narrator bitingly and ironically comments, "Doubtless there is more enlightenment now," and it is uncertain

> whether the farmers are all public-spirited, the shopkeepers nobly independent, the Sproxton men entirely sober and judicious, the Dissenters quite without narrowness of asperity in religion and politics, and the publicans all fit, like Gaius, to be the friends of an apostle. (epilogue 477)

Felix exists in the now, but his present culture is neither ready nor willing to embrace drastic reforms. Felix's desire for radical change promotes dissonance and disorder and further agitates an already hostile working class. What Felix's experiences of "we" as a part of a social collective teach him is that change must be made in increments, and it must be gradual, rather than precipitant.

Felix is not able to regulate the behavior and attitudes of men toward drastic moral and political change. He realizes the laborious and methodical process of assiduously inspiring social transformation by enlightening one voter at a time. The radical counterculture Felix incited provided him only a fleeting opportunity to reinvent himself in opposition to traditional values, needs, and identities that must be harmonized to maintain social order. As Felix leaves Treby Magna, he is aware that he has planted the seeds of social progress, seeds that will eventually grow to fruition with the passage of the Second Reform Bill in 1867. Thus, at the end of the novel, one radical is transformed and consumed by the dominant culture, while others are taking ideological root in the past.

Fertile soils of cultural and psychological landscapes will enable the growth of an emergent Victorian middle class. The new middle class will revive the class attitudes and habitus of past dominant bourgeoisie. The new middle class will repopulate the old ranks of the social hierarchy it once challenged. The new middle class will reinscribe place-identity through new modes of discourse that borrow from past conceptions of public and private space, conceptions that are continuously reconstituted to serve the interests of those belonging to the present culture, to those who are always living in the Now.

Chapter 5

Conclusion

During the period 1849–1866, novelists such as Brontë, Dickens, and Eliot considered how technologies of production, facilitated in the early to mid-nineteenth century, began altering the ways in which linguistic exchanges were employed by Victorians to occupy and negotiate public space. Rather than acknowledging the mutual reciprocity essential to linguistic exchanges, each novelist either directly or indirectly suggests how technologies fostered a new, unidirectional mode of speech that conditioned readers to accept their limited contribution to the metanarrative of print culture, which stabilized the status quo by assuring ideological conformity. As print culture created a community of readers, it also created a docile pool of supplicants who would fulfill the requirements of mechanized labor. However, within print culture, novelists like Dickens, Brontë, and Eliot emerged and challenged the ways in which printed texts were to be conceptualized within the context of larger social problems such as the inadequate representation of the poor and working classes, the deficiencies of educational systems, the volatility of factory environments, and the stratification of classes and genders.

Within the context of *Hard Times*, *Shirley*, and *Felix Holt*, characters either affirm or resist social categories such as class and gender. The characters demonstrate how alterations in linguistic exchanges, precipitated by the expansion of technologies within the early to mid-Victorian periods, diminished the opportunity for reciprocal transferences of power within the public sphere, transferences that may have yielded greater dispositions of social equilibrium. In *Hard Times*, Dickens provides readers glimpses of a community enveloped by industrialization, so much so that a new technical language of factuality is proposed to restrict and to limit the possibilities of achieving physical or ideological movement through differentiation. Through rigid systems of learning, factuality is promoted by Coketown's provisional leaders, men like Bounderby and Gradgrind, who administer technical principles of expediency and efficiency not only to the factory workers but also to the children who populate the ranks of Coketown's classrooms. The utilitarian philosophies of Coketown's provisional leaders underscore the prohibitive nature of a new system of rationality predicated on technical principles. As these principles become standardized and embedded in language and culture, they are reproduced and replicated within social institutions that seek to assure normative values and order.

The selected novels render critique possible by locating critical speech within the narrative form. By tracing the presence of speech as techné for invigorating responses to the social problems the novels report, the novelists revive speech as a mode of resistance to technologies of production as well as to mass literacy. Both seek to diffuse moral accountability by distancing us from the consequences of our actions, even if those actions entail willful compliance with the status quo. By experiencing distantiation, both the factory worker and the passive reader are made susceptible to modes of domination. By altering their proximity to the choices they make each day, technologies of production and print culture allow workers and readers to initiate actions and to view or to conceptualize the repercussions of those actions at a distance, far removed from consequences. Technologies of production and print culture devalue critique by divesting workers and readers alike of moral certitude. Without some sense of obligation to assure and to promote human agency, individual responsibility is displaced by operative terms which reorganize and redistribute human resources across an emerging technical sphere of engagement.

Brontë's *Shirley* uses the Luddite riots of 1811 to 1812 against the suboptimization of human labor to express how the radical sentiments of the Luddites empowered crowd formations. As Brontë's characters experience their assimilation within crowd formations, they also choose their degree of involvement. In so doing, they recognize the capabilities, interests, and collective will of crowd formations and suggest the development of a public sphere of engagement in which individuals struggle to express singularity but also embrace plurality of movement within crowds.

Finally, in Eliot's *Felix Holt*, readers find the radicalized individual, displaced from the crowd and struggling to locate a new sociospatial identity. Eliot offers readers the radical character of Felix Holt to suggest the ways in which conceptions of morality and class necessitate performative acts, which cause the inscription of sociospatial identity. It is this process of inscribing identity which determines one's proximity to historically prescribed definitions of belonging that have been used to assign placement within social hierarchies.

In each of the selected novels, cultures of exchange enable characters to create among themselves common expectations and interests and the capabilities for achieving their fulfillment. Within each exchange, linkages and bonds between individuals and groups are engendered. In each of these fictional cultures, characters facilitate modes of accommodation, conciliation, and consensus through overlapping linkages and bonds that isolate and minimize the opportunity for initiating critique by encouraging them to accept less than they wish for or desire. The fictional societies of the industrial novels replicate and represent the hierarchical society of a government organized by task-based commands. It is through task-based commands that all social members are perceived as belonging to the same moral and behavioral worlds. These task-based commands restrict the possibility of a spontaneous society and become unidirectional modes of discourse that condition social behavior.

"Read," "reflect," "consider," and "discuss" are the prescriptive commands and controls inherent in a work of fiction. These are the task-based commands that regulate readership of fiction, and these commands become even more poignant and more prohibitive in industrial novels. These commands condition readers by engaging them in a more active participation in the discussion of the novel. Through organized discussions, readers conceptualize accountability and become more motivated and responsive to their assigned roles within the group. The pockets of resistance that do form are treated as anomalies and are isolated to the periphery of society. These isolated groupings are allowed to exist because their integration within larger social orders authenticates unequal power exchange as a means to legitimate coercion and domination by neglecting intellectual and altruistic exchanges.

Technologies of production have been enabling the redesign of a plurality of worlds in which humans operate. Technical rationality realizes these locations within objective, subjective, and social worlds that are each being transformed by the fulfillment of technical principles, principles that legitimate the loss of critique through efficiency of information. The degree to which technical thinking has been affecting the ability to express self-awareness is incalculable. Technology no longer exists as merely a subsystem of government, culture, or society, nor does it function solely as a means for serving specific political or economic interests. Rather, it is the metaphorical "well" from which all knowledge springs forth onto a virtual plane of language, sense, and mind, extending consciousness beyond the physical body. It is Creation left incomplete by the absence of its human creator. It is a world characterized by the absence of disagreement, the absence of *différance*. It is a world no longer able to trace the presence of critique.

Notes

Chapter 1

1. All citations from the selected novels will be referenced by either book or volume, chapter, and page number.
2. As Belsey suggests, "the displacement of subjectivity across a range of discourses implies a range of positions from which the subject grasps itself and its relations with the real, and these positions may be incompatible or contradictory. These incompatibilities and contradictions within what is taken for granted . . . exert a pressure on concrete individuals to seek new, non-contradictory subject positions." See *Critical Practice* 65. Subjectivity is expressed through various modes of discourse, yet it is through these modes of discourse that we adjust and reinvest values and meanings to satisfy common conceptions of political and social worlds.
3. Simon Schaeffer quotes from Charles Babbage's essay *The Exposition of 1851*, see 171. In 1834 Babbage developed the principle of the analytic engine, which has been considered a precursor of the modern computer.
4. John Picker's *Victorian Soundscapes* reflects on novels like Eliot's *Deronda* from Victorians' aural experience. Picker attempts to situate himself within "the growing body of recent scholarship on nineteenth-century photography, optics, and visual culture" (6). Picker's concept of close-listening builds upon the work done in other musicological studies of nineteenth-century Britain by examining the aural experience of Victorian listeners and the adjustment to modern subjectivity to intersecting modes of technology, art, culture, and science. Although intriguing, Picker's study provides only a brief comment on Dickens' *Hard Times* (32), no comment whatsoever on Brontë or her novel *Shirley*, and but a short quotation from the "Author's Introduction" of the novel *Felix Holt*, explaining how we might similarly isolate modes of silence in Eliot's *Daniel Deronda* (92–93). Picker also

fails to mention how Eliot's reflections might better relate to her seminal work, *Middlemarch*, which itself explains, "if we had a keen vision and feeling of all ordinary human life, it would be like hearing the grass grow and the squirrel's heart beat, and we should die of that roar which lies on the other side of silence" (20.192). This excerpt from *Middlemarch* further substantiates the excess of signification in symbols that operate as signifiers of spoken discourse. Concepts like violence, crowd formations, and symbolic representations of counterculture were bespeaking new modes of social discourse, modes of discourse which were not limited to verbal speech but made use of symbols as a means of expression, while still performing the function of speech in the Victorian period.

5. See Guy.
6. *Formal* or *technical rationality* is predicated upon efficiency and expediency of production that verifies "means" and "ends" without favoring concerns for the public good. Technical rationality and formal rationality should be considered synonymous within the following discussion. *Substantive rationality* continuously adjusts the means to assure the public good. See Weber. See also Lee and Newby 175. As David Lee and Howard Newby suggest, formal rationality ". . . is the result of calculation." However, one must question whether or not such calculations have been imposing limits on the possibilities of engaging human thought and action within the technological sphere. Speech as techné enables movement away from the continuous flow of calculations and technical principles.
7. Habermas describes the lifeworld as the common interests, values, and routines we imbue with meaning in order to shape our lives. See "Technology and Science as 'Ideology'" 117; see also *The Theory of Communicative Action* 135–137.
8. Aristotle defines *techné* as those things which are caused by human arts or crafting. For the purposes of this discussion, speech will be employed as a techné for fulfilling the performative act of sublimation that releases humans from the constraints of technical thinking. Employing public speech as a requirement of membership within the public sphere allows for the possibility that

communicative rationality might reinvigorate critique within the context of the selected industrial novels. However, the supradisciplinary approach to technology makes this claim somewhat problematic, as it has become entwined in the moral, economic, and political fabric of postmodern society. See *Physics* 192b.

9. See Kaplan. Saul Kripke's supplementary essay "Speaker's Reference and Semantic Reference" offers a critical response to Kaplan's theories on referential expression.

Chapter 2

10. In Latin, the quotation reads as follows: "Nam qui dabat olim/imperium, fasces, legiones, omnia, nunc se/continet atque duas tantum res anxius optat,/panem et circenses." See Juvenal lines 78–91.
11. See especially Stallybrass and White's introduction and chapter 3, "The City, the Sewer, the Gaze and the Contaminating Touch." Stallybrass and White suggest that "the creation of a sublimated public body without smells, without coarse laughter, without organs [, or uncontrollable openings,] separate from . . . the ['grotesque' proletarian body of the] market square, alehouse, and street . . . was the great Labour of bourgeois culture" (93–94).
12. In chapter 3, "Addressing the Subject," Belsey argues that realistic novels offer contradiction but fail to "foreground" such contradictions in terms that challenge the status quo. These types of novels yield to a "hierarchy of discourses" which place contradiction in opposition to a more positivist perspective which attempts to reconcile them.
13. I am borrowing terms from the title of Stanley Diamond's essay "The Rule of Law Versus the Order of Custom."

Chapter 3

14. In his article "The 'Machine Breakers' and the Industrial Revolution," Alessandro Nuvolari eloquently summarizes this view of working-class resistance to mechanized labor.

15. Although Plotz does note the shape-shifting of crowds, he does not adequately emphasize the ways in which linguistic power exchanges enable its permutations and augment its volatility. As crowds transform themselves, they absorb and retain power, so much so that even as individuals remove themselves or are purged from their formation, they attract new members who more adequately support their mutating composition.
16. Bourdieu's conception of symbolic violence is considered throughout much of his works, including *Outline of a Theory of Practice*. Bourdieu defined a number of terminologies, such as cultural, economic, social, and symbolic capital, as well as concepts of habitus, field, and symbolic violence, many of which are employed within the critical framework of this chapter. See also his "On Symbolic Power" 163–170, *Masculine Domination*, and "The Forms of Capital" 241–258.
17. Reicher and Hopkins, *Self and Nation*, as well as Reicher and Hopkins, "On the Science of the Art of Leadership" 197–209.
18. Ian Munro suggests crowds in early modern London provided the "infinite connectability" between genders and classes. See Munro 132.
19. See Lyndall Gordon's biographical commentary on Brontë's "fears of her own misanthropy" (239).
20. Plotz's book *The Crowd: British Literature and Public Politics* builds upon this critical framework by analyzing and evaluating the emergence of the crowd as a literary trope in six industrial novels dated between 1800 and 1850, including Book Seven of William Wordsworth's *The Prelude* (1805), Maria Edgeworth's *Harrington* (1817), Thomas De Quincey's *Confessions of an English Opium-Eater* (1821), "The English Mailcoach" (1849), Thomas Carlyle's *Chartism* (1839), and Charlotte Brontë's *Shirley* (1849). Plotz reflects upon "the effects of these new crowds, riots, and demonstrations on the period's literature" (2). Plotz argues that literary texts of the nineteenth century recorded "features of the era's crowds that no other historical sources can supply" and suggests how the crowds became the novelists' "potent rival to the representational claims [of literature]" (2). In "Chapter 6: Producing Privacy in Public," Plotz considers the ways in which industri-

alization altered the "role of literature itself within . . . discursive space" as well as British conceptions of how one occupies public and private domains by representing "the crowd" as a polymorphous embodiment of society (2).

21. Tammy Clewell examines the ways in which Toni Morrison responds to African American history by placing her fictions at the center of discussions about mourning and identity. Her study reflects well on the idea of role-playing and enabled my own theoretical framework to take shape. See Clewell 131.
22. See Lisa Rodensky's talk from the Conference on Victorian Anxieties.

Chapter 4

23. See Ernst Bloch's definition of supplementarity in "Nonsynchronism and the Obligation to Its Dialectics" 22.
24. I will be citing chapter and page number for parenthetical references.
25. The following titles are insightful contributions to a growing corpus of readings on memory, trace, and presence-absence. See Sutton; Bolzoni 16–26; Fivush and Reese 115–132; McCauley 126–144; McClelland 69–90.
26. See Sommerfeldt 247. As Christina Sommerfeldt explains, Eliot uses literary references to align her novel more closely with classic epics and tragedies: "There are also allusions to *The Aeneid*, Dante's *Inferno*, and to Ovid's *Metamorphoses* (which of course is not epic, but suggests Eliot's alliance with high culture" (247).
27. Introduced within the House of Commons in March of 1831 and composed by then–Whig Prime Minister Charles Grey, Second Earl Grey, the First Reform Bill attempted to reconcile the disparities in representation between rural areas populated by upper-class nobility and the emerging middle classes of industrial cities of England. However, the legislative act failed to assure proportional representation for the lower class and redistributed representation to satisfy the demands of a politically vocal middle class. The Second Reform Bill of 1837 was initiated by Tory Benjamin Disraeli,

and it expanded the electorate by removing property ownership as a criteria for guaranteeing voting privileges.

28. Ruthrof analyzes and evaluates various "orthodox" theories of meaning from Gottlob Frege at the beginning of the twentieth century to theorists of the postmodern period before evidencing his own theory of "corporeal semantics," which suggests that the meaning of linguistic expressions is not discernable without the body as a visual aid of expression (190).
29. Horowitz 132. Horowitz's essay was reduced for inclusion in Pangallo's edited work. The original essay was published in *Texas Studies in Literature and Language* 17 (1975).
30. Ruthrof remembers Charles Sanders Peirce's notion of an "infinite semiosis" and suggests its importance by reconsidering Umberto Eco's comparison of semiosis to the "workings of a dictionary." Ruthrof praises Peirce's "overall semiotics, which accounts for everything thinkable in terms of inter-connected signs" (190).
31. See Pangallo 238. Karen L. Pangallo comments on the ways in which Eliot's novel becomes a "vehicle for promoting her ethos of duty and self-renunciation" (209).
32. Eliot speaks through the narrator, describing this "interconnectedness" as the "mutual influence of dissimilar destinies" (3.50).
33. Lukács restates Marx's conception of economic categories as "forms of existence, determinants of existence," which define the "material of economic categories philosophically" while also offering a "key to the literary portrayal of economic determination." See Lukács 294.
34. Human geographers Y. F. Tuan and D. Gutting have considered the ways in which linguistic traditions like storytelling affect the creation of place-identities. See Gutting 482–490. Also, see Tuan 684–696.
35. As Derrida suggests, "The supplementary of difference vicariously stands in for presence due to its primordial self-deficiency." Thus, the supplement signifies an "in-betweeness" of identity and nonidentity, of existence and nonexistence that expresses its "inherent lack." Derrida discusses the philosophy of language and the interpretations of signs in relation to logic and rhetoric. He critiques Edmund Husserl's claims that language is founded on

logic rather than on rhetoric and further suggests that meaningful language is limited to expression. See *Speech and Phenomena*, 88.

36. Lukács explains the concept of the "dramatic collision": "for it is clear that the socio-historical concentrating of contradictions in life necessarily demands a dramatic embodiment. . . . The contradictoriness of social development, the intensifications of these considerations to the point of tragic collision *general* fact of life" (99).

Bibliography

Adorno, Theodor. "On Lyric Poetry and Society." Trans. Sherry Weber Nicholsen. *Notes to Literature*. Vol. 1. Ed. Roif Tiedemann. New York: Columbia University Press, 1991. 37–54. Print.

Arendt, Hannah. *The Human Condition*. Chicago: University of Chicago Press, 1998. Print.

Aristotle. *Nicomachean Ethics*. New York: Pearson, 1962. Print.

———. *Physics*. Lincoln: University of Nebraska Press, 1961. Print.

———. *Rhetoric*. New York: Wilder Press, 2009. Print.

Armstrong, Nancy. *Desire and Domestic Fiction: A Political History of the Novel*. New York: Oxford University Press, 1987. Print.

Arnold, Matthew. *Culture and Anarchy and Other Writings*. Ed. Stefan Collini. Cambridge: Cambridge University Press, 1993. Print.

Babbage, Charles. *The Exposition of 1851*. 2nd ed. London: John Murray, 1851. Print.

Badiou, Alain. *L'Ethique*. Paris: Hatier, 1993. Print.

———. *L'Etre et l'Evénement*. Paris: Seuil, 1988. Print.

Bakhtin, Mikhail. *Rabelais and His World*. Trans. Hélène Iswolsky. Bloomington: Indiana University Press, 1993. Print.

Bammer, Angelica, ed. *Displacements: Cultural Identities in Question*. Indianapolis: Indiana University Press, 2001. Print.

Belsey, Catherine. *Critical Practice*. London: Methuen, 1980. Print.

Berger, Peter L., and Thomas Luckmann. *The Sacred Canopy*. New York: Anchor, 1990. Print.

———. *The Social Construction of Reality: A Treatise in the Sociology of Knowledge*. Garden City, NY: Doubleday, 1966. Print.

Bhabha, Homi K. "DissemiNation: Time, Narrative, and the Margins of the Modern Nation." *The Location of Culture*. New York: Routledge, 1994. 139–170. Print.

Bloch, Ernst. "Nonsynchronism and the Obligation to Its Dialectics." Trans. Mark Ritter. *New German Critique* 11 (spring 1977): 22–38. Print.

Bolzoni, Linda. "The Play of Images: The Art of Memory from Its Origins to the Seventeenth Century." *The Enchanted Loom: Chapters in the History of Neuroscience*. Ed. P. Corsi. Oxford: Oxford University Press, 1991. 16–26. Print.

Borgmann, Albert. *Holding onto Reality: The Nature of Information at the Turn of the Millennium*. Chicago: University of Chicago Press, 1999. Print.

Bourdieu, Pierre. *Ce que parler veut dire: l'économie des échanges linguistiques*. Paris: Fayard, 1982. Print.

———. "The Forms of Capital." *Handbook of Theory and Research for the Sociology of Education*. Ed. J. G. Richardson. New York: Greenwood, 1986. 241–258. Print.

———. *Masculine Domination.* Cambridge, U.K.: Polity Press, 2001. Print.

———."On Symbolic Power." *Language and Symbolic Power.* Ed. J. B. Thompson. Trans. G. Raymond and M. Adamson. Cambridge, U.K.: Polity Press, 1991. 163–170. Print.

———. *Outline of a Theory of Practice.* Trans. Robert Nice. Cambridge: Cambridge University Press, 1972. Print.

Brontë, Charlotte. *Shirley.* Eds. Herbert Rosengarten and Margaret Smith. Oxford: Oxford University Press, 1998. Print.

Burnett, John. *The Annals of Labour: Autobiographies of British Working Class People, 1820–1920.* Bloomington: Indiana University Press, 1974. Print.

Carroll, Joseph. *Evolution and Literary Theory.* Columbia: University of Missouri Press, 1995. Print.

Clewell, Tammy. "From Destructive to Constructive Haunting in Toni Morrison's *Paradise.*" *West Coast Line* 37 (spring 2002): 131–142. Print.

Connor, Steven. "Deconstructing Hard Times." Web. 27 Oct. 2006. <http://www.andrew.cmu/edu/course/76-101qt/DeconstructingHardTimes.txt>.

Darwin, Charles. *The Origin of Species by Means of Natural Selection.* University of Missouri St. Louis. Web. 11 Nov. 2006. <http://www.umsl.edu/~engjcarr/web_documents/darwin%20edition%20introduc.htm>.

Deleuze, Gilles, and Claire Parnet. *Dialogues.* Paris: Flammarion, 1977. Print.

Derrida, Jacques. *Of Grammatology*. Trans. Gayatri Chakravorty Spivak. Baltimore: Johns Hopkins University Press, 1976. Print.

———. *Specters of Marx*. Trans. Peggy Kamuf. New York: Routledge, 1994. Print.

———. *Speech and Phenomena*. Trans. D. B. Allison. Evanston, IL: Northwestern University Press, 1990. Print.

———. *Writing and Difference*. Trans. Alan Bass. Chicago: University of Chicago Press, 1978. Print.

Diamond, Stanley. "The Rule of Law Versus the Order of Custom." *In Search of the Primitive*. New Brunswick, NJ: Transaction, 1981. 255–280. Print.

Dickens, Charles. *Hard Times*. Eds. Fred Kaplan and Sylvère Monod. New York: Norton, 2001. Print.

Dixon, John, and Kevin Durrheim. "Displacing Place-Identity: A Discursive Approach to Locating Self and Other." *British Journal of Social Psychology* 39 (2000): 27–44. Print.

Durvall, Frank O. *Popular Disturbances and Public Order in Regency England*. London: Oxford University Press, 1969. Print.

Eliot, George. *Felix Holt: The Radical*. 1896. Ed. Lynda Mugglestone. New York: Penguin, 1995. Print.

———. *Middlemarch*. 1871. Ed. Rosemary Ashton. New York: Penguin, 1994. Print.

Ermarth, Elizabeth D. *Sequel to History: Postmodernism and the Crisis of Representational Time*. Princeton, NJ: Princeton University Press, 1992. Print.

Fivush, Robyn, and Elaine Reese. "The Social Construction of Autobiographical Memory." *Theoretical Perspectives on Autobiographical Memory*. Ed. M. Conway. New York: Kluwer, 1992. 115–132. Print.

Fluck, Winifreid. "The Americanization of Modern Culture." *Cultural Policy, or the Politics of Culture? The Proceedings of the Seventh International Conference of the Polish Association for American Studies, November 1997*. Eds. Agata Preis-Smith and Piotr Skurowski. Warsaw: Institute of English Studies, University of Warsaw, 1999. 17–49. Print.

Foucault, Michel. *Language, Counter-Memory, Practice*. Ed. Donald F. Bouchard. Ithaca, NY: Cornell University Press, 1977. Print.

———. "The Life of Infamous Men." *Michel Foucault: Power, Truth, Strategy*. Eds. M. Morris and P. Patton. Sydney: Feral, 1979. 76–91. Print.

———. "Of Other Spaces." *Diacritics* 16.1 (spring 1986): 22–27. Print.

———. "What Is Enlightenment?" *The Foucault Reader*. Ed. Paul Rainbow. New York: Pantheon, 1984. 32–50. Print.

Galloway, Shirley. "The Casuistry of George Eliot." *LitList*. 1993. Web. 11 Nov. 2006. <http://www.cyberpat.com/shirlsite/essays/casuist.html>.

Gardiner, Michael. "Bakhtin's Carnival: Utopia as Critique." *Utopian Studies* 3.2 (1992): 21–50. Print.

Gissing, George. *The Nether World*. London: Dent, Everyman Library, 1973. Print.

Gordon, Lyndall. *Charlotte Brontë: A Passionate Life*. New York: Norton, 1995. Print.

Gutting, D. "Narrative Identity and Residential History." *Area* 28 (1996): 482–490. Print.

Guy, Josephine M. *Victorian Social-Problem Novel: The Market, the Individual and Communal Life*. Hampshire, U.K.: Palgrave, 1996. Print.

Habermas, Jürgen. "An Alternative Way out of the Philosophy of the Subject of Communicative versus Subjective-Centered Reason." Trans. Frederick G. Lawrence. *The Philosophical Discourse of Modernity: Twelve Lectures*. Cambridge, MA: MIT, 1996. 294–326. Print.

———. "A Reply." *Communicative Action: Essays on Jürgen Habermas's "The Theory of Communicative Action."* Ed. Axel Honneth and Hans Joas. Boston: MIT, 1991. 214–264. Print.

———. "Technology and Science as 'Ideology.'" *Toward a Rational Society*. Trans. Jeremy Shapiro. Boston: Beacon, 1970. 81–127. Print.

———. *The Theory of Communicative Action. Volume Two: Lifeworld and System: A Critique of Functionalist Reason*. Trans. Thomas McCarthy. Boston: Beacon, 1987. Print.

Heaney, Seamus. *Seeing Things*. London: Faber, 1991. Print.

Heidegger, Martin. "The Question Concerning Technology." *The Question Concerning Technology, and Other Essays*. Trans. William Lovitt. New York: Harper, 1977. 3–35. Print.

———. "Traditional Language and Technological Language." Trans. W. Gregory. *Journal of Philosophical Research* 23: 129–145. Print.

———. *Überlieferte Sprache und Technische Sprache*. St. Gallen: Erker, 1989. Print.

Horowitz, Lenore W. "George Eliot's Vision of Society in Felix Holt, the Radical." *The Critical Response to George Eliot*. Ed. Karen L. Pangallo. Westport, CT: Greenwood, 1994. 132–145. Print.

Hutcheon, Linda. *A Theory of Parody: The Teachings of Twentieth-Century Art Forms*. New York: Methuen, 1985. Print.

"Idea of Literacy." 8 Dec. 2006. Web. <www.espach.salford.ac.uk/english/suplimentarymaterial/opp_supplement.htm.>

Jones, R. T. *George Eliot*. Cambridge: Cambridge University Press, 1970. Print.

Juvenal [D. Ivnii Ivvenalis]. *Satirae X*. Edidit breuique apparatu critico instruxit. Web. <http://www.curculio.org/Juvenal/s10.html>.

Kaplan, David. "Demonstratives: An Essay on the Semantics, Logic, Metaphysics, and Epistemology of Demonstratives and Other Indexicals." *Themes from Kaplan*. Ed. Joseph Almog, John Perry, and Howard Wettstein. New York: Oxford University Press, 1989. 481–563. Print.

Kellner, Douglas. "Techno-Politics, New Technologies, and the New Public Spheres." *Illuminations* (December 1997): n. pag. Web. 11 Nov. 2006. <http://www.uta.edu/huma/illuminations/kell32.htm>.

Kreilkamp, Ivan. *Voice and the Victorian Storyteller*. Ed. Gillian Beer. Cambridge: Cambridge University Press, 2005. Print.

Kripke, Saul. "Speaker's Reference and Semantic Reference." *Contemporary Perspectives in the Philosophy of Language*. Ed. Theodore Uehling and Peter French Jr. Minneapolis: University of Minnesota Press, 1977. 6–27. Print.

Kucich, John. *Repression in Victorian Fiction: Charlotte Brontë, George Eliot, and Charles Dickens*. Berkeley: University of California Press, 1987. Print.

Lane, Christopher. "Charlotte Brontë on the Pleasure of Hating." *ELH* 69.1 (2002): 199–222. Print.

LeBon, Gustave. *The Crowd: A Study of the Popular Mind*. 1895. Introd. by Robert A. Nye. New Brunswick, NJ: Transaction, 1995. Print.

Lee, David, and Howard Newby. *The Problem of Sociology: An Introduction to the Discipline*. New York: Routledge, 1990. Print.

Lukács, Georg. *The Historical Novel*. 1962. Trans. Hannah Mitchell and Stanley Mitchell. Lincoln: University of Nebraska Press, 1983. Print.

Marcuse, Hebert. *One-Dimensional Man*. Boston: Beacon, 1964. Print.

———. "Some Social Implications of Modern Technology." *The Essential Frankfurt School Reader*. Ed. Andrew Arato and Eike Gebhardt. New York, Continuum, 1982. 138–182. Print.

McCauley, Robert N. "Walking in Our Own Footsteps: Autobiographical Memory and Reconstruction." *Remembering Recon-*

sidered: Ecological and Traditional Approaches to the Study of Memory. Ed. U. Neisser and E. Winograd. Cambridge: Cambridge University Press, 1988. 126–144. Print.

McClelland, James L. "Constructive Memory and Memory Distortions: A Parallel-Distributed Processing Approach." *Memory Distortion: How Minds, Brains, and Societies Reconstruct the Past*. Ed. D. L. Schacter. Cambridge, MA: Harvard University Press, 1995. 69–90. Print.

McCuskey, Brian. "Not at Home: Servants, Scholars, and the Uncanny." *PMLA* 121.2 (1 March 2006): 421–436. Print.

Messenger, Nigel. "'The Dung-Heap and the Flower': Gissing's *Nether World*." Web. 12 Dec. 2006. <http://www.mfo.ac.uk/Publications/actes1/messenger.htm>.

Miller, Angela. "Everywhere and Nowhere: The Making of the National Landscape." *American Literary History* 4.2 (1992): 207–229. Print.

Morson, Gary S., and Caryl Emerson. *Mikhail Bakhtin: Creation of a Prosaics*. Stanford, CA: Stanford University Press, 1990. Print.

Munro, Ian. *The Figure of the Crowd in Early Modern London: The City and Its Double*. New York: Palgrave, 2005. Print.

Nietzsche, Friedrich. *Basic Writings of Nietzsche*. Trans. and ed. Waller Kaufmann. New York: Modern Library, 1991. Print.

———. *Thus Spoke Zarathustra: A Book for Everyone and No One*. 1883–1885. Trans. R. J. Hollingdale. Harmondsworth, U.K.: Penguin, 1969. Print.

Nuvolari, Allesandro. "The 'Machine Breakers' and the Industrial Revolution." *Journal of European Economic History* 31.2 (2002): 393–426. Print.

Pangallo, Karen L., ed. *The Critical Response to George Eliot.* Westport, CT: Greenwood, 1994. Print.

Paris, Bernard J. *Experiments in Life: George Eliot's Quest for Values*. Detroit, MI: Wayne State University Press, 1965. Print.

Picker, John M. *Victorian Soundscapes*. New York: Oxford University Press, 2003. Print.

Plotz, John. *The Crowd: British Literature and Public Politics*. Berkeley: University of California Press, 2000. Print.

Postmes, Thomas. *Deindividuation*. Web. 18 Aug. 2006. <http://www.ex.ac.uk/~tpostmes/deindividuation.html>.

Reicher, Stephen, and Nick Hopkins. "On the Science of the Art of Leadership." *Identity, Leadership and Power.* Ed. D. van Knippenberg and M. Hogg. Oxford: Blackwell, 2003. 197–209. Print.

———. *Self and Nation*. London: Sage, 2001. Print.

Richardson, J. G., ed. *Handbook of Theory and Research for the Sociology of Education*. New York: Greenwood, 1986. Print.

Rodensky, Lisa. "Novel Anxieties: Barnaby Rudge and Victorian Crowd Control." Conference on Victorian Anxieties. University of Hartford. 11–13 April 2002. Address.

Rowles, G. D. "Place and Personal Identity in Old Age: Observations from Appalachia." *Journal of Environmental Psychology* 3 (1983): 299–313. Print.

Ruthrof, Horst. *Semantics and the Body: Meaning from Frege to the Postmodern.* London: University of Toronto Press, 1997. Print.

Schaeffer, Jason. "Harlequin Britain: Pantomime and Entertainment, 1690-1760 (review)." *Theatre Journal* 57.2 (May 2005): 341–342. Print.

Sommerfeldt, Christina. "The Landscape, the Coach-Journey, and the Claims of High-Culture Narrative: The Introduction to George Eliot's *Felix Holt*." *Culture and the State: Landscape & Ecology*. Ed. James Gifford and Gabrielle Zezulka-Mailloux. New York: Routledge, 1997. 243–249. Print.

Stallybrass, Peter, and Allon White. *The Politics and Poetics of Transgression*. Ithaca, NY: Cornell University Press, 1986. Print.

Sutton, John. *Philosophy and Memory Traces: Descartes to Connectionism*. Cambridge: Cambridge University Press, 1998. Print.

Taneja, Leena. "Consciousness, Literature and the Arts." *Archive* 5.2 (August 2004): n. pag. Web. 11 Nov. 2006.<http://www.aber.ac.uk/tfts/journal/archive/taneja.html>.

Thompson, E. P. "The Moral Economy of the English Crowd in the Eighteenth Century." *Past and Present* 50 (Feb. 1971): 76–136. Print.

Tuan, Y. F. "Language and the Making of Place: A Narrative-Descriptive Approach." *Annals of the Association of American Geographers* 81 (1991): 684–696. Print.

Watt. Ian P. *The Rise of the Novel: Studies in Defoe, Richardson and Fielding*. 1957. Berkeley: University of California Press, 2001. Print.

Weber, Max. *The Protestant Ethic and the Spirit of Capitalism*. 1930. Trans. Talcott Parsons. New York: Routledge, 1992. Print.

Index

Made in the USA
Monee, IL
12 November 2026

d0845a98-8448-4b28-9b7d-75a4fb0a057aR01